*This book is dedicated
to all ocean entrepreneurs
and the innovations they
have yet to discover*

The
new
fish
wave

The
new
fish
wave

Leete's Island Books

Library of Congress Number:
2019914549
ISBN: 978-0-918172-78-5
Copyright © 2020 Thor Sigfusson

Design and illustrations:
Arnar & Arnar | arnarogarnar.com

Typeface in headings:
Harbor by Krot & Krass

Portait Photograps:
Eva Lind Gígja

Photo on page 101:
Ernir Eyjólfsson
Photo on page 21:
Sigurgeir Jónasson
Photo on pages 57 and 97 :
The Ocean Cluster

Leete's Island Books,
Box 1 Sedgwick, Maine 04676

First edition

—

leetesislandbooks.com

Leete's Island Books

Contents

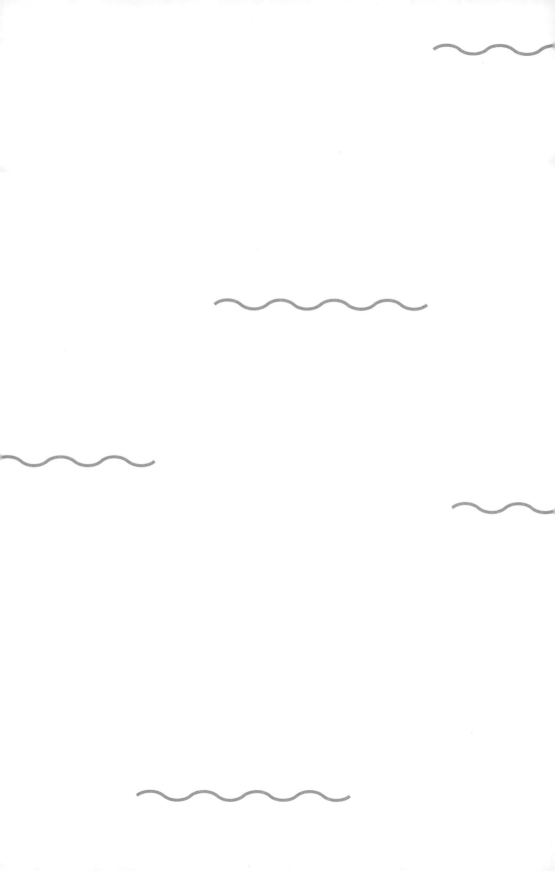

14 LIFE BELOW
WATER

GOAL #14

—

Conserve & sustainably use the oceans, seas and marine resources

The world's oceans — their temperature, chemistry, currents and life — drive global systems that make the Earth habitable for humankind. Our rainwater, drinking water, weather, climate, coastlines, much of our food, and even the oxygen in the air we breathe are all ultimately provided and regulated by the sea. Throughout history, oceans and seas have been vital conduits for trade and transportation.

Careful management of this essential global resource is a key feature of a sustainable future. However, at the current time, there is a continuous deterioration of coastal waters owing to pollution and ocean acidification. It is having an adverse effect on the functioning of ecosystems and biodiversity. This is also negatively impacting small-scale fisheries.

Preface:
The New
Fish Wave

The world can learn from Iceland, the small fishing nation in the North Atlantic, which has, in many ways, transformed itself from being one of the poorest countries in the world a century ago to one of the richest nations in the world today. By showing pride in its seafood industry and using innovation to safeguard the environment, the nation has created wealth, derived more value from each fish and managed fisheries in a sustainable way. Iceland has now become a niche leader in fisheries.

This book will describe how an industry cluster can be an agent of change for any industry, using Iceland's seafood industry as an example. The Iceland Ocean Cluster (IOC), which started as research at the University of Iceland, has become a vital part of the "new" seafood industry in Iceland where cross-pollination between old wisdom and new knowledge is vital to rejuvenating a traditional, nature-based field.

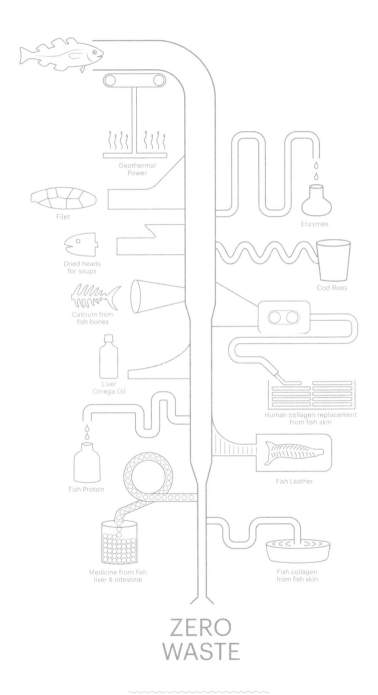

Geothermal
Power

Filet

Dried heads
for soups

Calcium from
fish bones

Liver
Omega Oil

Fish Protein

Medicine from fish
liver & intestine

Enzymes

Cod Roes

Human collagen replacement
from fish skin

Fish Leather

Fish collagen
from fish skin

ZERO
WASTE

*The incredible fish value machine describes the mission of
the Iceland Ocean Cluster: to use resources to the fullest.*

The initial study at the University of Iceland showed that tech entrepreneurs in the seafood industry were communicating among themselves in ways which were different to other industries. While entrepreneurs in other industries formed diffuse networks with more contacts, the seafood techs had smaller, closed networks. We wondered why we couldn't forge more of these kinds of diffuse networks in natural resource industries, such as fisheries, giving the entrepreneurs more opportunities to connect with a wider network and broader backgrounds. The IOC was initiated with this goal in mind. We now understand that when many companies come together in the "new" seafood industry, they're greater than the sum of their parts, and they can create something new. The IOC, which was formed in 2011, has been able to use Iceland's national focus on seafood to create a vibrant startup community. It has revolutionized the existing seafood industry and encouraged its involvement in the growing cluster movement. The IOC is a catalyst of change for Icelandic seafood: a new fish wave!

This book gives readers a step-by-step guide toward a successful seafood cluster, using the example of the IOC. Even though our initiative was named "cluster", the ideas presented here are not limited to forming a cluster. Other kinds of networks and organizations can ignite dynamic networking of people and ideas, like we did, through sharing their experiences and comparing their perspectives.

This book lists the five steps which I see as crucial in establishing an ocean cluster. These steps are:

» It's all about economics

An in-depth economic study of the whole seafood sector gives a clearer picture of the "real" importance of the seafood industry. An economic study can also open the door for further discussions about the growth potential in this industry.

» Mapping it all

Cluster mapping is a crucial step in the formation of a cluster. What are the main companies, major supporting services and R&D institutions — and how do they all fit together? Mapping provides valuable information about the industry and the concentration of economic activities.

» Finding the Leaders

An aspiring cluster founder — anyone who is most interested in setting up ocean clusters in their own country or region — needs to be particularly mindful of their leadership team. An ideal team would be comprised of dynamic group of individuals who are engaged in the developement of the cluster.

» Low hanging fruits

If business leaders are to be active in the cluster, we need to keep them busy with interesting projects and 'low hanging fruits.' I use 'low hanging fruit' here to refer to the simple tasks that can be completed first, creating positive results.

» Strategy is key

Getting people to meet will always be the core activity of the cluster. At the same time, all meetings and "low-hanging fruit" achievements need to be driven by long-term strategy. A clear, game-changing mission for an industry cluster is absolutely essential. In our cluster, we decided on a program we call "100% Fish".

The New Fish Wave is just emerging globally. I see it in growing awareness of the importance of healthy oceans and also in greater understanding of the health benefits of the natural ocean proteins. I see it also in the interest many fishing nations have in doing more with less: to stop discarding large parts of the fish. Clusters have an important role to play in this new movement: to connect people with often very different backgrounds and skills to make way for further product development and value creation. Our greatest success will probably be where we connect veteran fishermen with R&D people who have never been on board a fishing vessel!

Doing more with less

The global seafood industry dumps nearly 10 million tons of perfectly good fish back into the ocean or uses it as landfill. At the same time, 33% of fish stocks are threatened by overfishing. Mentioning fisheries to most developed countries seldom creates excitement. Many Western countries are somehow embarrassed about their fisheries; most of them know their fisheries are discarding huge amounts of fish, but little is being done.

This is hard for Icelanders to understand, since Icelandic fisheries have been using more of each fish than most Western countries — making a profit from many parts of the fish which other countries dump into landfills.

For over a hundred years, the Icelandic fishing sector has proven itself to be the most productive fisheries sector in the world. As a result, Icelandic fishermen and the entire seafood industry have continuously been making an effort to stay the best.

Since the 9th century, Icelanders have derived vitality and stamina from fish. Seafarers dropped hand lines into the sea, caught fish, gutted and then hung them to dry on driftwood racks. Sea trousers softened with fish oil allowed fishermen to stay warm, dry, and go out further away from the shore.

Wooden rowboats led to sailing smacks and motor driven trawlers which reached even further into the North Atlantic swells. The fishing crafts may have changed, but the Icelandic determination to push the limits of what is possible remains constant.

To this day, fisheries remain one of the pillars of the Icelandic economy. However, like many other countries, Iceland has faced reduced landings and has been mindful not to overfish. Since the early 1980s, Iceland has enforced an efficient fisheries management system which assigns fishermen a specific quota for each fish stock based on the TAC (Total Allowable Catch) determined by the government. In many ways this system as well as an overall mindset of utilising each fish to the fullest, have laid the ground work for the success of the modern fishing industry in Iceland where the attitude has more and more become "lets do more with less". The graph below compares the value of the Icelandic cod catch in 1981 and 2018. In 1981, Icelanders caught 460 thousand tons with a turnover of US 1095 million. By 2018, the catch had dropped significantly down to 252 thousand tons with a turnover of US 924 million . The drop in catch was 45% but the value dropped only by just under 16% (US fixed price). Here, we have only taken into account the export value of the cod fillets.

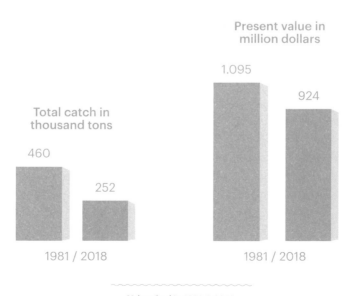

Value of cod in 1981 & 2018.

Icelanders have shown it is possible to significantly increase the value of fish catch. Through various forms of innovation, Iceland gets more value from each fish than many other developed fishing countries. The difference is even higher between Iceland and various developing countries. There are lots of opportunities to do better with the wild fish protein, since it is one of the best natural proteins in the world.

In Icelandic, *nýtin* is a positive word that describes a person who uses things to their fullest. Of course, the positive attitude towards more utilisation in Iceland is partly explained by the fact that Icelanders are not rich with resources — so they must treat the resources they have with dignity. We rely so heavily on fisheries that we cannot afford to treat this resource lightly.

Innovative health, pharmaceutical and even fashion products from wild fish stock are being developed in the Icelandic economy. There lies our opportunity: Icelanders are putting their minds to work, creating more value from each fish. The results are already out: We are getting at least 30% more value from each cod than most developed countries. Fish is not only the fillet, but it is also fish skin becoming health products, the liver becoming omega oils and pharmaceuticals, the head and bones used in various ways — basically, nothing is left for landfill.

This island of 360,000 inhabitants now has over 70 tech firms selling technology in the global market. Marel, one of the largest providers of food processing technology in the world, leads the pack. Marel was founded by young engineers at the University of Iceland who began to experiment with electronics and early computers. Marel is a major contributor to the ongoing technological revolution in the global fish processing industry. A processing plant in Iceland with an output of 1000 tons would have around 80 employees in 1990, 40 employees in 2010 and most likely 10 employees in 2020. While the manual labour required for production decreases with the implementation of sensible cluster strategy, new jobs are created by seafood tech and seafood biotech enterprises, many yet to be imagined.

Iceland has also over 50 firms specialising in various byproducts and processing of seafood. Lýsi, a global producer of refined fish oils for human consumption, is one of the leaders in byproduct utilisation in Iceland. Both

Marel and Lýsi have benefitted from being a part of a society that emphasizes creating more value from limited resources.

Almost anybody can do less with more. On the other hand, doing more with less requires a lot of "people skills" and the right motivation. In a scarce natural resource industry, doing more with less is not only a neccessity, it is a great opportunity.

By setting our minds on doing more with less, we start to get a better understanding of the oceans' amazing protein offerings. These proteins enable the world to live more healthily. Doing more with less can enrich the world — especially coastal towns, which are thought to have been left behind by the 21st century economy.

Did the Icelandic wave start wIth an ITQ System?

Probably the greatest game changer in the Icelandic seafood industry was the introduction of the Individual Transferable Quota system (ITQ), implemented in steps from 1976 to 2004. This system is the cornerstone of the fisheries system in Iceland and the key ingredient for the successful resource management in Iceland. In other words, it limits the total catch and ensures that catches are in line with the total allowable catch (TAC).

Dr. Ragnar Árnason professor at the University of Iceland:
"There is evidence of substantially improved resource stewardship under the ITQ system. First, TACs are now generally adhered to. Second, and more importantly, there are pretty clear indications that the fishing industry, i.e. the holders of ITQs, are much more willing now than before to accept and even support radical reductions in TACs in order to rebuild fish stocks".

The early days

The Westman Islands, my birthplace, is a grouping of islands south of Iceland. The biggest island, Heimaey, has over four thousand inhabitants. For the last centuries, the Westman Islands have been the crown prince of fisheries in Iceland. The islands enjoy rich fishing grounds all around them. There has always been an honor culture on the island: fishermen are kings. People never ate cod — it was like eating American dollar bills. Instead, we exported our valuable cod and consumed cheaper white fish.

It would be absolutely great for the buildup of the story to describe my ancestors as great seafarers and fishermen, however they were not. My grandparents moved to the island in the '30s where my grandfather would become a schoolmaster. In those days, fishermen were not overly excited about education and would rather like to see the children busy at work in the fish processing factory. This became a long and hard battle between education and work in the small fishing village. Education won!

The hard working culture did not disappear even though the children stayed at school. Summer jobs for the young were plentiful. I was fortunate enough to get a summer job processing fish in my childhood. My task was to layer cod fish for export in containers with sea salt. The job didn't really inspire, but when I received my pay cheque, I was rich and I felt I could do anything with my life. Pay cheques for 14 year olds were a great way to make young people feel worthy.

In 1983, my brother Dr. Þorsteinn I Sigfússon, later professor at the University of Iceland and CEO of Iceland Innovation Center, challenged me to write a short article about biotechnology. He suggested the Icelandic title "Gull úr gor" which translates "Gold from guts". I was nineteen years old and had very little interest in biotechnology. The article was published but received no attention. I did not pursue this topic further for some years. However, I am sure this short piece and my brother's enthusiasm sparked a light which later led to my enthusiasm for the new seafood industry.

Years later, in the year 2002, I decided to pursue a Ph.D. degree. At this time I had written two books about the small state of Iceland in the global economy and global expansion of Icelandic companies. I had also written a prize-winning essay which I called the "Knowledge Network Economy". This was in the '90s, the early days of the internet, and I "radically" predicted that as distance mattered less in modern times, Iceland could become a global hub for niche products and services. Some of my predictions look awfully silly today, but I see now that I was already becoming a network nerd and a firm believer in strengthening business through relationship networks.

My plan was to focus my Ph.D. research on how entrepreneurs use their network and past experience to expand abroad. Of course for Icelandic innovators, living in a community of three hundred and fifty thousand people, expanding abroad is not only an option, it's a necessity. My initial research focused on whether a business founder's background could explain different approaches and attitudes towards globalizing their businesses. We found significant differences between founders who had studied abroad and those who had not; the former had a much stronger appetite for expanding abroad.

In a similar study two years later, I decided to focus only on startups and their founders to find out what might explain their different approaches and attitudes towards globalisation. I specifically looked for founders of tech companies who had engineering or technical background. This work became the primary reason for the establishment of the IOC. At this point, there was still no fishery focus in my work. However, some of the tech entrepreneurs I studied had been working closely with the seafood industry.
I was amazed to learn how different the networking strategy had been among the people I interviewed. Some had very large personal networks and used them frequently. Others had very small networks and relied heavily on a few strong contacts. The difference didn't seem to be explained by global or domestic education, founders' characteristics, etc. Was there something else?

I soon discovered that the seafood tech entrepreneurs seemed to be less connected overall, and that they utilised less of their relationship network within the industry compared to other entrepreneurs in different sectors. The seafood tech entrepreneurs had also smaller networks outside their home region. This idea inspired me to study relationship networks in my Ph.D. and focus on relationships of entrepreneurs.

*The hard working culture did not
disappear even though the children stayed at school.*

The weak ties, the strong ties, & social capital

The central thesis of social capital theory is that "relationships matter" (Field, 2003) and it would later become a crucial part of the ideology behind the IOC. These "relationships" are knitted in social networks which are a valuable asset enabling people to commit themselves to each other. Network relationships generate social capital (Arenius, 2002), a sense of belonging, and the concrete experience of social networks (and the relationships of trust and tolerance that can be involved) can, it is argued, bring great benefits to people (Powell & Grodal, 2005).

Entrepreneurs require resources such as capital, skills, and labor to start or expand their business activities. The entrepreneurs hold some of these resources themselves, but they access other resources by using their contacts and professional network. This "group effort" is central to raising both creativity and efficiency. Social capital is the outcome of successful contacts and is the key component of entrepreneurial networks (Burt, 1992). Maurer and Ebers (2006) define social capital as "an asset available to individual or collective actors that draws on these actors' positions in a social network and/or the content of these actors' social relations" (p. 262). Networks are also important in identifying opportunities (Johanson & Vahlne, 2006) which is a crucial part of a cluster's success.

In 1973, Mark Granovetter came up with an idea in sociology which would later have a huge influence on the study of relationships in networks. The idea was fairly simple: weak relationship ties can act as important bridges in a network building, and for that purpose, they are sometimes more important than strong relationship ties (family and friends). Granovetter's research showed that when people think about who might help them in a job search, they tend to make a short list of very close friends and family — the strong ties. However, Granovetter's research showed that your closest friends are not really your best bet when searching for a job because their network is very much like your own. So, to get to a larger group of people in your job search, you are better off tying up with people with whom you have weak ties (people you barely know) rather than those with whom you have strong ties. Granovetter suggested that strong network ties have high levels of social relationship or personal interaction with high frequency. The strong network ties mean that members are motivated to be of assistance and protect actors in insecure positions. Weak ties, however, are not as heavily based on personal interaction among members of the network but may provide a strategic advantage in terms of resource availability. Weak ties act as "local bridges" to parts of the network which would otherwise be disconnected and they also offer new opportunities.

Being linked to a network is not the only issue; one must be in a position to do something within that network. There, the cluster idea appears. How can clusters assist in building weak and strong ties among individuals? Here, I will emphasise personal relationships rather than organisational relationships. In my experience, it's all about people and trust. You will rarely find organisations with high long-term trust which enables them to work together. The underlying trust is between individuals in these organisations.

Industry clusters

The success of clusters is explained by the social relations among community members. Through clusters, networks of contacts emerge between individuals across normally firm boundaries and act as channels, flowing knowledge and business opportunities among themselves.

The concept of clusters is generally traced back to Alfred Marshall's *Principles of Economics*, which was first published in 1890. According to Marshall, agglomeration advantages were supposed to be linked to three sets of localisation economies: a pooled market for workers; availability of specialised inputs and services; and technology spillovers.

Contemporary work on industry clusters is heavily influenced by ideas of agglomeration theorists such as those that have already been mentioned. In general, these ideas focus on external economies of scale, industrial linkages, and the factors that give economic advantages to individual firms located close to other similar or related firms. However, after the late 1980s, a different kind of work emerged. It calls for a more detailed analysis of industry agglomeration, in which the economics are more embedded with social and cultural aspects. Pyke (1990) introduced the term "New Industrial Districts" to describe those districts that were characterised by smaller scale production, flexibility, specialisation, and more cooperation between contracting firms to make vertically disintegrated production possible. This concept is often referred to as the "flexible specialisation school" in which the main emphasis is on social and cultural networks with interpersonal relations, on face-to-face encounters, and on casual and informal flow of information. This knowledge sharing built on trust and informal relations is a good business.

With more emphasis on knowledge-based economies, the relationship between industrial agglomeration and competitiveness is to a greater extent explained by enlisting local knowledge, rather than analysing external economies of scale or natural advantage. The network of actors (i.e. firms and institutions) who enjoy geographical proximity and shared cultural, linguistic and social norms and values, determines the innovative performance of companies, their growth, and their competitiveness. Innovation is supposed to be best where there are high levels of interaction and face-to-

DR. ÁGÚSTA GUÐMUNDSDÓTTIR OF ZYMETECH

*Zymetech ehf. was founded in 1999 by Ágústa and her husband, Dr. Jón Bragi
Bjarnason, both professors at the University of Iceland. They gained deep understanding
of the efficacy of enzymes from deep sea cod. Zymetech extracts the enzymes from parts of
the cod viscera that would otherwise be discarded.*

face contacts. According to this school of thought, the cost of transmitting knowledge increases with ever-increasing distance.

My research focus was mainly on relationship networks of seafood-technology entrepreneurs and industry clusters. The theoretical foundation of my studies became a combination of network and cluster theory. Michael Porter (1990), one of the principal pioneers of cluster research in the world, defines clusters in economic sectors as:

> *...a group of related companies, suppliers, service providers, companies in related sectors and public bodies... in specialised fields that compete among themselves but also work together.*

Cluster analysis has been used to explain the development of industries and business in specific regions. It has subsequently been used to explain the growth and development of towns, cities and even nations. This analysis has also shown that regional industries emerge where there is a strong cluster environment.

The developmental history is typically as follows: progress of the economy and/or technology in a certain region creates favorable economic conditions. These conditions are often based on natural resources, although this is not a prerequisite. Individual companies begin operations in the area and form a base industry. A group of companies in related fields then form around this base industry. They are linked to each other in numerous ways and support each other: for example, this environment may promote the development of human resources, technology, and technological equipment which benefit them all. The companies, moreover, circulate assets among themselves through transactions and transfers of employees, and at the same time, create a demand for further services and infrastructure. Cluster analysis involves a detailed examination of such developments, and the manner in which the cluster's industries are connected and the arrangement of their manufacturing processes.

Cluster's role in
~ relationship networks ~

This model below presents a suggested view on the role of clusters in strengthening entrepreneurs' business relationship network. According to the model, the individual entrepreneur sets the tone for the relationship opportunities in their network. It also shows how an industry cluster can affect relationship networks.

Part A of the model shows how entrepreneurs' individual skill sets may affect their relationship networks. The model recognizes different entrepreneurs and their different levels of relationship skills.

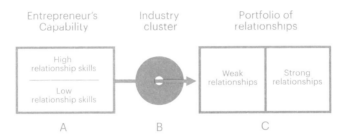

Entrepreneur's Capability	Industry cluster	Portfolio of relationships
High relationship skills / Low relationship skills		Weak relationships / Strong relationships
A	B	C

In **Part B** of the model, the cluster assists the entrepreneurs to build their networks. This model shows that industry cluster preference affects the "process through which ties are selected for entrepreneurship" (Jack, 2010: 133).

The relationship resources in **Part C** shows the portfolio of the entrepreneur's relationships, both weak and strong. Here, the cluster can affect the portfolio and increase the entrepreneur's network.

Relation-
ships among
tech entre-
preneurs

My earliest research into relationships and clusters focused on relationship networks among tech entrepreneurs. In my first preliminary interviews for my Ph.D. research, conducted in 2010, I interviewed five tech entrepreneurs. Two of them were involved with seafood technology. Through these interviews, I stumbled upon an interesting finding: The two seafood-technology entrepreneurs had a smaller relationship network than the others, even though all of them were actively seeking international business opportunities and had similar educational backgrounds.

The most likely explanation why the relationship networks seem to be smaller in businesses related to seafood is that the seafood industry is a natural resource industry. I have later come to suspect that, where natural resources are scarce, entrepreneurs are more likely to keep the resources for themselves and have a fairly small network of people they trust.

My first real-world test of Granovetter's theory of weak and strong ties was to bring together seafood-technology entrepreneurs from different parts of the seafood value chain in Iceland. In late 2010, I gathered a group of 20 seafood-technology entrepreneurs to discuss their business and whether their group could gain from further relationships. This group had never had a formal platform to meet. One-third of these firms are members of the Federation of Industry, but the Federation had not formed any special group for the marine technology firms.

This meeting deepened my drive to study the seafood-technology entrepreneurs' relationship network, since it became apparent the people involved barely knew each other. Most of them had met at the annual Fisheries Exhibition in Brussels, but they hadn't formed further bonds. We knew there were great opportunities for cooperation among these entrepreneurs for marketing, innovation and the exchange of good practices.

My next step was to examine the relationship networks of individual seafood-technology entrepreneurs. For this, I studied entrepreneurs who owned small firms (fewer than ten employees at the time of research), focusing on the relationships among the entrepreneurs and the nature of these relationships. The entrepreneurs were all founder-owners (or part owners) responsible for general management.

After the initial networking events in Iceland, where I witnessed seafood tech entrepreneurs introducing themselves to each other for the very first time, I interviewed the entrepreneurs. I asked why they had not met before, given that they could learn a great deal from each other and collaborate on projects important to each of them.

Most of them responded by saying they didn't have time for "socialising." I will always remember one entrepreneur who had doubts about my interpretation. He said:

I have very strong business links with great customers in local fisheries and I nurture them. With others, such as these tech colleagues, I knew about their existence. If I needed to contact them, I would just do so. We are so few on this island, we know everybody!

My findings indicated that most seafood-tech entrepreneurs had developed strong relationships with one or two fishing companies, and that they were able to obtain knowledge and early revenues from these colleagues fairly easily. The fishing companies, for their part, quickly saw value in the entrepreneurs' skills and knowledge. Therefore, it was natural for seafood-tech entrepreneurs to focus on these domestic relationships, who were, as one entrepreneur noted:

… These contacts were… in our home garden and we could get in touch with them right here.

This led to entrepreneurs relying on their strong domestic relationships in every aspect of their business development. Still, most of the entrepreneurs had, from the start, planned for their products to reach a global market.

A picture of these entrepreneurs' networks began to develop. They had already commercialised through domestic relationships that being local, were easily developed. They then sought global relationships, but this was a new and unfamiliar task for these entrepreneurs. Some of the domestic relationships, however, were part of global networks and could fulfill the role of 'introducers' (Coviello, 2006). For example, one entrepreneur said:

We had high hopes that the part ownership of the Icelandic fish processing plant by a Belgian firm would push us further into the European market with our technology.

Most research indicates that in the early stages, most entrepreneurs would seek to rely on strong trust and knowledge-based relationships. They seek people who can add directly to the competitive advantage of their new venture, including those who may help introduce and/or develop new customer relationships. All the seafood-technology entrepreneurs had developed their products in relationships with individual domestic firms and people they knew very well. This effect, in the early days, was profound:

> *I got a kick-start as they (an Icelandic company) bought my concept and idea from day one.*

This early establishment of strong relationships with local companies considerably benefited the development and establishment of the entrepreneurs' young firms.

First, it helped the firms to attract other resources. These domestic relationships presented valuable possibilities for developing connections in other sectors and territories. Most importantly, the focus at this stage was building relations with potential business clients, irrespective of territory, which, for the seafood-technology entrepreneurs meant:

> *… finding the customer who is willing to try our technology.*

The strong relationships the seafood-technology entrepreneurs established early on presented them with a ready source of clients that could be further developed. The strong relationships the seafood-technology entrepreneurs established early on presented them with a ready source of clients that could be further developed, giving them an almost-instant revenue stream.

All the entrepreneurs I studied worked hard to develop strong domestic relationships, and in some cases their local partnerships even helped them go global:

> *A part owner of the second Icelandic processing firm we worked with is a foreign investor. Our first contract abroad was with these guys and it was realised through the Icelandic firm.*

Second, having a proven product which was domestically successful made it much easier to generate global interest, especially at trade fairs.

Third, they inherited a worldwide strategic network identity based on Icelandic prominence in the seafood industry. Several discussed the benefits gained from the domestic track record and having strong Icelandic specialists in fisheries and seafood.

While forming strong bonds with a few domestic clients worked in the beginning, it may have led them to be less strategic in their overall network building. There are two main reasons for this. First, even though the entrepreneur's domestic relationships often led to initial exports, they rarely led to strong global relationships. The domestic client industry was inwardly focused, so maintaining strong relations with domestic relationships did not greatly help the entrepreneurs to develop a strong international network. Second, the entrepreneurs were so reliant on their strong local connections that they became hesitant to form new relationships or strengthen weak ones. One participant illustrated the approach well:

> I have a Scottish client whom I met at a trade show years ago. We have developed very good relations which has led to further product development.

In these interviews, I noted the seafood-technology entrepreneurs had a relatively protective attitude towards relationships. They had strong network relationships with a few buyer companies, and they worked hard to nurture and solidify these bonds. It was natural for them to protect their valuable assets:

> There is a lot of duplication going on. An employee of mine left a couple of years ago … with my invention, just changing the look but using similar ideas and software.

The island mentality

I n general, my findings indicated that these entrepreneurs regarded their networks with an "island mentality," in which you think you know everybody and can connect to them whenever needed. The problem with this attitude is that those connections, in reality, rarely occur. It is crucial for an entrepreneur to be good at extending both their domestic and global networks in order to develop a sustainable business. Connecting with people outside of their local market is vitally important, particularly if that market is very small. Many of the entrepreneurs I spoke with did not nurture their weak ties and were instead happily focusing on the few strong ties that upheld the status quo and allowed their business to merely survive. Could a cluster extend the network of entrepreneurs and inspire them to actively use this network to grow their business?

The general conclusion from these first interviews was that, prior to the establishment of this cluster, tech entrepreneurs had not made significant effort to cooperate.

One seafood-technology entrepreneur responded this way:

> *If I need to cooperate I will just pick up the phone and make*
> *a call.*

When asked how much he had done that in the past 12 months, he reflected:

> *I don't recall. I have not had any particular needs to do so in*
> *some time.*

This island mentality is the main reason why many firms pay attention only to their closest associates rather than a possible extended network. It may also be the reason why many parts of the seafood industry are dominated by small, family-owned enterprises. These companies lack the ability or desire to work together in groups, instead trusting very few people, mainly family members.

Building a cluster can increase the level of trust among small firms. Higher trust level increases collaboration among the firms and their larger relationship network strengthens their competitiveness. There are scale economies in trusted networks of small firms. The network allows the small firms the luxury of being small but having the strength, with their network, to compete with much larger firms.

Soon after the initial group meeting with the twenty CEOs, the group met again. The meeting strongly underlined the group's interest in networking, talking and collaborating.

One participant reflected:

> *I did not know that we were so many. I think it could be an*
> *idea for us to see if we can cooperate at trade shows. Most of*
> *us are small and it takes a lot of effort for my small operation*
> *to participate as a stand-alone in a trade show.*

At the next meeting, the participants met at the sites of two of the participant's operations.

This gave the group a good opportunity to start building knowledge of and trust in each other.

One entrepreneur noted:

> *I see various co-op possibilities regarding environmental issues, marketing, technology exchange, communication etc.*
>
> *We might consider cooperating on the development of small boats where Icelandic seafood-technology firms have had a competitive advantage.*

Soon after I had established the group, the goal became clear: to broaden their horizons and further build relationships.

From natural- to knowledge resources

Unlike natural resources which are limited, the more you use knowledge resources, the faster they grow. A traditional natural resource industry will always have tendencies towards an island mentality: you keep your limited resources for yourself. As the "new seafood" industry moves up the value pyramid, from limited natural resources to more knowledge-based resources, the mindset must change. The companies in "new seafood" are knowledge-based and, as such, must rely much more on their ability to network and innovate. Knowledge needs to adapt to a changing environment, and companies need to be continuously sharing ideas.

A company like Kerecis in Iceland, which creates medical skin supplements from cod skin, is a good example. Kerecis' proprietary scientific process increases the value of the cod skin they use from 20 cents on the dollar to over 2500 dollars per skin. The cross pollination of ideas and knowledge is the piece which the seafood industry needs to add to the puzzle to take the step further into a more dynamic knowledge-based industry.

THE VALUE PYRAMID OF FISH PRODUCTS

The value of seafood created through the scientific process can be increased by 1000%.

Through my studies of the relationships in the seafood industry, I realised a clear opportunity to connect people. Still, I needed more groundwork and data to have a clear picture of the size of the whole seafood sector in Iceland. That was to become my first major work.

It's all about economics

Looking at the seafood through the eyes of an economist, incredible team effort appears: the fisherman, the machinist, the engineer, the processor, the biologist, the chemist, the salesman, the retailer, the consumer… There are many actors who play a crucial role a crucial role, though many have never met. There is no one person in charge of the whole value chain. This is what makes an industry so fascinating.

In this chapter, I will describe how our economic research led to some of the most extensive data available on a seafood value chain in an economy. This research was very important for the development of the IOC. Our research opened many eyes. We were able to clearly lay out the complex interaction among all of the actors in the value chain and show how each of them played an important role. This research paved the way for the game changing role which the cluster has played in the seafood industry in Iceland.

I knew that the IOC had to have a strong basis in data in order for it to be respected. Our work had to be well grounded and substantial. My background is in economics, so I was determined that, prior to the formation of the IOC, the nation should get a crash course in the economics of the modern seafood industry in Iceland.

I had read a lot through the years about industry clusters, especially renowned clusters such as the tech hub in Silicon Valley, the wind energy movement in Denmark and the salmon farming industry in Norway. I had been fortunate to be invited to Silicon Valley in California by the U.S. Government in 1996 and was overwhelmed. We did not meet any specific cluster managers in the Valley, but I felt the spirit and pride was there. This visit was one of the reasons why I grew to believe that Iceland could become the Silicon Valley of our niche industries.

I started reading literature on industry clusters such as the *Cluster Development Handbook* by Ifor Ffowcs-Williams. Ifor had been a frequent guest speaker in Iceland on industry clusters, and I saw his simple road map of successful clusters as a convenient method to try. Ifor emphasises the importance of good groundwork before starting an industry cluster, and I saw this also as an important step in the cluster development.

Fisheries are, in essence, a natural resource industry — a so called base industry — in which a wide range of companies service this natural resource sector or add value to its products. These companies are not perceived as a part of the seafood industry in national statistics.

The underestimated base industry

Dr. Ragnar Árnason, a professor at the University of Iceland, and I published a paper in 2012, where we attempted to analyze the nature and scope of the whole seafood industry over time, evaluate its future growth potentials, and describe the conditions needed in order for the industry to grow. This had not been done in a detailed way prior to our study.

This paper and the research collaboration with Dr. Árnason became one of the pillars of the IOC. I believe we set standards for future research which could make it easier to compare performance between years. It would allow us to further examine the cluster's growth potentials and the conditions that must be in place to utilise these potentials. With the right nurturing, the seafood industry could become even more productive to the broader economy.

Icelandic fisheries evolved from traditional fishing and fish processing into a diverse business sector consisting of fields such as technology, logistics, marketing and more. Prior to our study, however, nobody had understood the big picture interactions of these fields. (This cross-pollination of fields within the seafood industry is what I call an "ocean cluster"). This shortage of information about the "whole picture" hampered discussions on its impact and abilities. I had been to too many public meetings where speakers would admire the fishing industry while discounting its ability to play a significant role in the future of the Icelandic economy. The public debate often implies that the number of jobs created by fisheries are declining. However, this debate only focuses on fisheries and fish processing, important foundations of the cluster, but far from its only activities. At the same time, popular belief assumes that fisheries will become less important for the economy in the years to come, as fishing and fish processing require less manpower.

We realised that the limited lens used to analyze the seafood industry in Iceland (and probably in most countries) underestimates the impact of the whole ocean cluster in national statistics. By looking at the entire ocean cluster, we can create more enthusiasm for the seafood industry.

This narrow view of the seafood industry is dragging it further down — and not just in Iceland. It is very important to address this lack of information on a global scale — the seafood industry itself should pressure the government to view the seafood industry with a "wider" and more correct "lens" to give a better perspective on the importance of the industry. This clearer picture of the real impact of the fishing industry could also strengthen its position in national debate.

The economic analysis

We set out to explore the economic effects of fisheries in the Icelandic economy. This is crucial work for every industry cluster, especially for one with such an emphasis on tech. Most people only think the industry involves catching fish and selling them on the market, but the reality is much more complex.

Pioneers in the fish industry have long been leading the world in tech advances like extracting high-end fish oils and enzymes. If they could do that with limited support, what could we do if we created a movement of entrepreneurs working in this field? In order to find out, we needed to create excitement and understanding for the broad scope of this vital industry. Part of the problem was clerical: in national data, the development of fish oil and enzymes for retail or skincare was not categorised as the "seafood industry" but as "industry". This meant that data on fisheries was stuck in the past and gave a static picture of an otherwise dynamic industry.

Árnason's student in economics at the University of Iceland, Linda Björk Bryndísardóttir, did tremendous work for us in collecting detailed data from fisheries about their purchases of services from different parts of the economy. By collecting the data, we could find "a number" which would better reflect the importance of a base industry like the seafood industry in Iceland.

According to the national accounts, the direct contribution from fisheries and fish processing to the GDP had been 7–10% over the past few years before our study, employing around 8,600 people or approximately 5% of Iceland's workforce. Despite this paltry percentage, the fishing industry is still considered to be the foundation of Iceland's economy. It could be tempting to assume that the belief of fisheries' importance is outdated. But is that really the case? Do these statistics provide a realistic view of the importance of the fisheries sector in the Icelandic economy?

It has long been obvious that the fisheries industry is more vital to Iceland's economy than the national accounts suggest. Árnason and Agnarsson (2005) pointed out that the fisheries sector is a base industry sector and that its total contribution to GDP was higher than its direct contribution. They prepared a statistical assessment of these overall effects and found the more accurate number to be between 25 and 35% of GDP.

Our research added to this discussion about the true role of the seafood industry and put more effort into obtaining industry data. We trained our focus on the whole ocean cluster in Iceland (termed the "traditional fisheries sector") and all the manufacturing activities it supports, whether directly or indirectly. We included manufacturing operations which sprang up under the umbrella of traditional fisheries industry: we found many companies which started domestically with the support of the fishing industry, but subsequently developed to stand on their own feet (and may even have started their own exports.) These companies would not exist at present if it were not for the industry's support, so they merited inclusion in the cluster.

Examining the ocean cluster as a whole, a much clearer picture of the importance of the fisheries sector in the Icelandic economy comes into focus. This picture is far more holistic and accurate than the old model of examining only the traditional fisheries and fish processing. We must point out that relatively simple changes to operating arrangements in the traditional fisheries sector, such as employing contractors for more tasks (i.e. outsourcing), can significantly distort the view of the economic importance of the sector if one only studies the fisheries sector. These tasks may be quite substantial, such as offloading catches, maintenance work on fishing vessels and fish processing plants, as well as a variety of other services. When a fishing company decides to purchase such services from other companies instead of performing them themselves, official figures can indicate a decrease in the number of people employed by the fisheries sector when in fact no decrease has occurred.

Fisheries as a base industry

The term 'base industry' has been under development for quite some time. Its origin can be traced to the research carried out by the German economic historian Werner Sombart early in the 20th century and subsequently to developments in regional economics in the latter part of the same century. Regional economics divides economic industries into two sectors: base economic industries and the manufacturing and service sectors that exist "under" the base industries. This second sector owes its foundation, development, and very existence to the base sector.

Roy, Árnason, and Schrank (2009) set forth a definition of the term 'base industry', which is as follows:

> *The economic base is an industry or a collection of industries that is disproportionately important to a region's economy in the sense that other economic industries depend on the operation of the economic base, but not vice versa, at least not to the same extent.*

Imagine an unpopulated region that is rich in a natural resource, such as valuable minerals or fishing grounds. The technology and knowledge to use the resource profitably is available and, as a result, so is funding. A workforce moves in and an industry forms to utilize the natural resource to its fullest. This industry is therefore defined as the base industry. Following the establishment of the base industry, numerous other industries that serve the base industry and its employers may surface. Some of these industries may provide the base industry with resources that can be economically manufactured in the area. These economic operations are so-called "backward connection" of the base industry. Other industries may be established for the purpose of meeting the demands of employees for goods and services to the extent that such manufacturing is feasible in the area. Such industries may include various public services. These derived industries and service operations also need a workforce and resources. This creates further demand and more industries, and so on and so forth. Overall, the scope of such derived operations can become quite large when compared with the base industry. The scope is first and foremost dependent on local ability to meet the demand for goods and services that the base industry creates, directly and indirectly.

The core of the matter is that all these derived operations are created by the establishment of the base industry and rest on the foundations that it has created. Without the base industry, these other industries would never have appeared. If the base industry leaves, e.g. because the mine of valuable minerals runs out or fish stocks are destroyed, the derived industries are at great risk to fail, too, unless the people in the region are able to find and develop a new base industry.

If the base industry were for some reason to shrink it would create a chain reaction. For instance, a possible catch failure would result in less income for the base industry, leading to revenue cuts in the service sectors. This would make it necessary to reduce the number of employees. The service sector would subsequently be downsized, and some of the businesses would cease operations. Thereby, the economic downturn in the region would be greater than the downturn in the base industry itself. This is, in fact, one of the main characteristics of base industries — they have a knock-on effect on the economy. If, for example, a cinema ceases operation, this would not have the same type of effect. It is most likely that people's business with a cinema would simply be transferred to similar leisure pastimes, i.e. the closure would not have knock-on effects, at least not to the same extent.

It's clear that in Iceland the fishing industry is a base industry. Iceland has valuable fishing grounds that would be exploited even though little or no services were available from land. This can be seen from the way foreigners fished in Icelandic fishing grounds for centuries, right into the 20th century, because this exploitation was generally carried out without any significant services from Icelanders. The current operation of the fisheries sector in Iceland, however, has called for a wide range of derived operations in the country.

FIGURE I, on the opposite page, attempts to outline the whole ocean cluster as it appeared in our research.

Figure 1

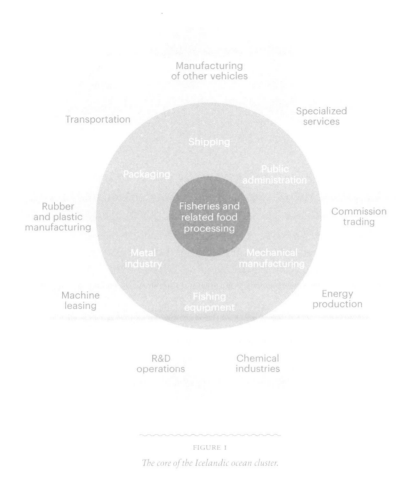

FIGURE 1

The core of the Icelandic ocean cluster.

The core of the ocean cluster and the central point in FIGURE 1 is the traditional fishing industry which consists of fisheries, fish processing, and marketing. This is the foundation or the base around which the cluster forms. As stated previously, only the fishing industry is the actual base industry as the processing and marketing of the products hinges on actually catching the fish.

In close connection with the fisheries sector is a group of industries that provide the sector with resources and services.

The industries form the inner circle around the fisheries sector in FIGURE 1. These include:

1. Packaging industry
2. Fishing gear manufacture
3. Shipping operations
4. Diverse mechanical manufacture
5. Metal industry
6. Public administration

These sectors emerged in the wake of the significant growth of the fisheries industry and have been developing and growing for decades. It is important to realise that the fisheries industry's demand for resources and services could have been met through importation and services from overseas. Luckily, there was enough initiative and production ability in Iceland to ensure the slow but steady growth of many of these service sectors locally.

Numerous industries are more loosely connected to the fisheries sector but should nevertheless be included in the ocean cluster, at least in part. These industries have been placed in the outer circle in FIGURE 1. They include:

1. The manufacture of rubber and plastic goods
2. Machine leasing
3. Energy production and utilities
4. Various research and development operations
5. Various types of chemical industries
6. Shipping/haulage operations
7. Commission trading
8. Various specialised services

It is important to realise that the industries in the inner and outer circle of FIGURE 1 are not only connected to the base industry, fisheries, but are also inter-connected among themselves, as well as being possibly connected to other industries outside the ocean cluster. Thus, for example, manufacturers of rubber and plastic goods enjoy benefits from the chemical industry, mechanical manufacturers and the metal industries. The same can be said of the fishing gear manufacturing industry, the chemical industry and the metal industry, which also support each other in various ways. All these are then connected to shipping services and specialised services, and so on and so forth. These connections are apparent in resources and products, but also in less tangible ways. For instance, specialist knowledge forms within one industry and flows between the others through information exchanges and trained, specialised employees. Thus all these companies and industries, listed in FIGURE 1, form an industry cluster, as defined by Porter, where each enjoys benefits from the others and the whole is stronger than each individual company. This forms the Icelandic ocean cluster.

Figure 2

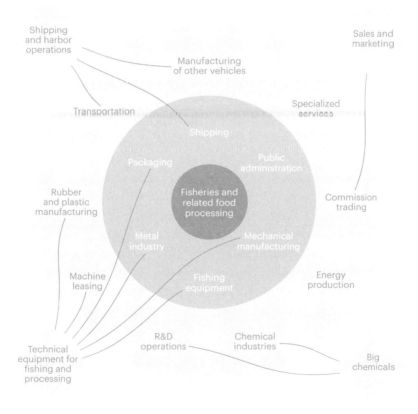

FIGURE 2

Examples of connections between industries in the ocean cluster.

FIGURE 2 explains the various relations between the industries in the ocean cluster. Note that, in addition to the connections shown in FIGURE 2, all the sectors are connected to the core in FIGURE 1: the base industry, fisheries. FIGURE 2 also indicates that the sectors in the cluster form sub-clusters. For instance, shipbuilding (which is a part of a sector named "Manufacture of other transport equipment" in the Statistics Iceland classification system), shipping and transport are industries that form a range of interactive connections and form a sub-cluster. Moreover, the metal industry, packaging industry, rubber and plastic goods manufacturers and fishing gear manufacturers have considerable internal connections, so these industries can be said to form a sub-cluster in connection with technology and equipment for processing and fishing. Within this sub-cluster are dozens of companies manufacturing goods for the fisheries industry, aquaculture industry, and others, which they offer on the international market.

The fisheries sector's contribution to GDP

The fisheries industry's contribution to the GDP can be divided into three parts.

1. Direct contribution, the added value that forms in the fishing industry itself.
2. Indirect contribution, the added value that forms in the industries that are responsible for supplying the fishing industry with resources (backward connections) or for further processing the industry's products (forward connections).
3. Induced impact, the added value that forms in sectors that provide the employees of the fishing industry and related industries (backward and forward connections) with goods and services.

Direct contribution

In our research, we found the total contribution of the fisheries sector (i.e. fishing and processing) to be approximately 10.2%; The contribution of fishing is 5.7% and processing 4.5%.

Indirect contribution

The fisheries sector's indirect contribution to the GDP is in the innovations and manufacturing by other firms in the ocean cluster, which trace their business dealings back to the fisheries sector. This indirect contribution is not calculated by Statistics Iceland. However, it is possible to estimate the contribution as a multiple of the added value that forms in the "side" industries (which Statistics Iceland does compile) and the share of the fisheries sector in their turnover. A prerequisite for this estimate is that the added value from business dealings with the fisheries sector is generally the same as in the industries' other business dealings.

This formula is based on extensive data acquisition. The study focused on a range of large and small companies in the fisheries sector, together controlling more than 20% of the total catch quota in Icelandic fishing waters. Researchers collected detailed information on all purchases these fisheries companies made from any other companies which could be considered part of Iceland's ocean cluster. These companies were then classified according to the company classification system used by Statistics Iceland. Thereafter well over a hundred companies in the company categories were contacted and information obtained on their turnover, human resources use and scope of operation in general. On this basis, it was then possible to estimate the total turnover in these sectors and thereby the share of their turnover that can be traced to business dealings with the traditional fisheries sector.

The fisheries industry's direct contribution to the GDP is due to industrial fishing and seafood processing. The fisheries sector's indirect contribution to the GDP was estimated in our research to be 7.3% of the GDP. The largest proportion of this contribution can be attributed to various types of services and industries outside the fisheries sector. The retail sector and the transport sector also represent substantial added value to the fisheries sector.

Consequently, the sum of direct and indirect added value, is approximately 17.5% of the GDP. Of that percentage, direct contribution is just under 60% and indirect just over 40%. I should reiterate that the indirect added value is relevant here because the fisheries sector is a base industry. Without the fisheries sector, the "side" industries would never have formed, and nothing would have replaced them.

Finally, I want to stress that these direct and indirect contributions are not the total contributions made by the fisheries sector to the GDP. We have yet to take account of the demand that the direct and indirect added value, traceable to the fisheries sector, creates in the economy and which is likely to encourage substantial production increases, as opposed to what would have been the case otherwise.

Induced impact

In addition to the direct and indirect contribution of the fisheries sector to the GDP, it stands to reason that the added value industries employ workers and generate profits, all of which will be used to purchase consumer goods and services. Thereby the sector has an even greater effect toward the increase of manufacturing in the economy. In our research, we chose to call such effects *induced impact*.

Here is an example to illustrate the concept. Added value is the sum of wages and profit. Let us imagine that an employee in the fisheries field (or related sector in the ocean cluster) receives a specific wage. He uses these wages to purchase goods and services and to pay public levies, and he keeps some as savings. Purchases of goods and services are equivalent to the demand in the markets in question. The proportion of this demand directed toward domestic goods and services encourages and creates conditions for more domestic manufacturing. The proportion of the added value that is spent on savings and public levies also leads to domestic demand, albeit indirectly. The state spends tax income in some manner to purchase goods and services. Savings are used, through the mediation of the financial system, for investments which also involve the purchase of goods and services. The same principles apply to this demand as to the demands of the wage earner. The proportion that is directed at domestic goods and services also creates conditions for increased domestic manufacture. This increased manufacture forms wages and profits for others, and so on and so forth. Thus, this *induced impact* leads to a chain reaction throughout the economy. When everything is taken into account, this induced impact can be quite

substantial. Investments, moreover, are by their nature, likely to increase the economy's manufacturing capacity and thereby lead to economic growth in the future.

No reliable investigations into the scope of this induced impact are available in Iceland. It is quite likely that they are between 50–100% of the direct and indirect contribution of the fisheries sector. This is in tune with available investigations into the total contribution of the fisheries sector to the GDP in Iceland (Agnarsson and Árnason, 2007). As previously stated, this direct and indirect impact was estimated to be 17.5% of the GDP. We estimated the induced impact to be approximately 7% of the GDP.

Living organism

Industry clusters tend to have some of the characteristics of a living organism. They can strengthen, grow in scope, and may even develop off-shoots that can establish themselves in a new industry. Thus, companies in the ocean cluster may expand to the extent that they can begin manufacturing for other industries, or even begin exporting independently of the Icelandic fisheries sector.

Thanks to the various methods of seafood byproduct utilisation, independent export turnover is estimated to amount to approximately 4% of the total exports from Iceland. It is unclear what this means in direct and indirect contributions to the GDP, i.e. added value. Conservatively estimated, however, this contribution could be between 1.5–2% of the GDP. Some of the firms (or even groups of firms) that come into being as a part of the base industry cluster may very well become independent, meaning that they establish trading partners outside the cluster and generate the majority of their revenues without the cluster. In some cases — like Iceland's seafood processing tech companies, for instance — they may even become the focal point of a new sub-cluster which forms around them. See FIGURE 3.

A wide range of companies connected with the fisheries sector has gradually developed in Iceland. These companies are responsible for supplying the sector with some of its resources and take its products for further processing and distribution. The scope of these operations is substantial. On the

Figure 3

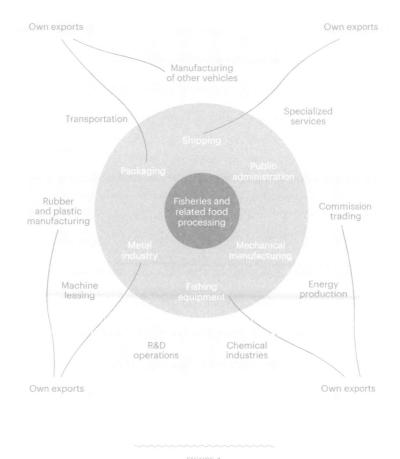

FIGURE 3

Independent export activities in the ocean cluster emerge.

scale of contribution to the GDP, it is almost as great as the direct contribution of the fisheries sector itself. It is important to keep in mind that these indirect contributions could not exist without a base industry, i.e. the fishing industry.

The fisheries sector and the related operations can overall be viewed as an industry cluster as defined by Michael Porter. The companies in this cluster are linked to one another in various ways and draw support from one another. The cluster, viewed as a whole, is therefore economically more dynamic, more efficient and more flexible than the simple sum of the companies that form it. The cluster operates, to some extent, as a very diverse company without the administrative disadvantages that characterize such companies. The Icelandic ocean cluster embodies Porter's definition of typical cluster effects.

There are several examples of this in Iceland's ocean cluster. Marel is probably the best known, but many others are up-and-coming, like Skaginn 3X, Valka, and Curio, manufacturers of fish processing technology, and Hampidjan, a fish gear technology company. However, there is absolutely no guarantee that industry clusters will grow and prosper. Clusters can also wither and die away. This can happen if the base industry, on which the cluster rests, suffers setbacks such as those due to altered operational conditions or less favourable competitive conditions.

It is crucial when forming a cluster (in this case, one involving the seafood industry) to properly present the seafood industry with reliable data and consequently get the industry out of the shadows! Unfortunately, these practices are rare in most countries. There is limited knowledge about the industry and, as mentioned previously, national statistics often do not paint it accurately. It is important to underscore that seafood is a "dynamic base industry" which can form the foundation for a diverse range of other industries. These other industries may become considerably larger than the initial base industry that spawned them.

There are numerous examples of this overseas. For instance, not long ago the Netherlands was a world leader in flower cultivation. Flower growing has now moved to other countries, while the Netherlands has become a leading force in flower marketing and sales.

In another example, many Finnish companies serviced Nokia with a diversity of operations related to mobile phones. Over time, they evolved and became global players in the field of mobile phone apps. The development of the Icelandic fisheries industry and the growth of diverse, related operations in the ocean cluster is merely one more example of the same phenom-

Fisheries industry	% of GDP
Direct contribution	10.20%
Indirect contribution	7.30%
Induced impact	7.00%
Other export operations	1.50%
Total	26.00%

TABLE I

Estimated contribution of the ocean cluster to the GDP.

enon. It shows how a dynamic base industry leads to a cluster of industries that multiply the base industry's contribution to the GDP. The cluster can even create a new, independent base industry.

In total, the direct and indirect contribution of the ocean cluster to the GDP, its induced impact, and the independent export operations it has fostered, equals approximately 26%. See TABLE I.

One of the main goals of this chapter was to explain how we have undergone a thorough study into the economic effects of the seafood industry in Iceland. By introducing such an in-depth study, we were able to convincingly show the "real" importance of the seafood industry. The voices perpetuating the popular notion of a fading seafood industry are taking note. It also opened the door for further discussions about the growth potential in this industry, and the many the opportunities within the ocean cluster in Iceland.

Mapping it all

Many coastal towns in Europe and North America had, in the past, plentiful jobs and opportunities as they benefited economically from resource-filled oceans. With significant cuts in fisheries, many of these areas have been stagnating for over 20 years.

Better linking these ecosystems to the startup/R&D world can change this pattern. The lack of networking among the entrepreneurs and cooperation among industries in the coastal areas may indicate there are opportunities which have not been utilised — not least opportunities for future generations of highly trained or well educated people. There are significant "blind spots" in large parts of these areas which have not been explored. Building seafood clusters in coastal communities with the aim of connecting different parts of the ecosystem can benefit everybody. In order to connect the parts, though, we must know what they are.

A cluster map is a crucial step in the formation of a cluster. Although the economic analysis of the cluster cannot be overstated as an important tool for identifying the cluster activity at a macro level, it is not sufficient. I realised I needed more qualitative information to provide me with the tools needed to start the cluster work. Thus, the next step was to obtain direct information through questioning the companies themselves. I asked, "what are the main companies, major supporting services and R&D institutions — and how do they all fit together?" I expected this hands-on approach to offer insight into the companies' characteristics and their managers' attitudes about relationships and collaboration. All told, I got a better idea of these enterprises' current projects and their plans for the future. This cluster mapping was became a very important tool for the establishment of the IOC. Mapping provides valuable information about the industry and the concentration of its economic activities. It is a key ingredient in assessing clusters.

A team of students worked with me, partially sponsored by a student fund. This was possible because we were mapping a small country; regardless, I highly recommend the "hands-on" method for anyone, especially if they are focused on a fairly small industry (like the seafood industry) or ocean-related industries in a particular region.

I knew from the beginning that good official data was available on the number of boats, number of fishing companies, fish processing companies, etc. It made our lives much easier. The challenge and the fun part was to map the rest: companies which have never been officially linked to the seafood sector; tech companies; various service companies; byproduct companies; and, of course, R&D, Universities, etc.

We obtained a list of all service providers to some of the largest fishing companies in Iceland. The initial work was to categorise these companies into different fields. Suddenly we had a good picture of the broad industry which was servicing the seafood industry. Now, the challenge was to get more data from these companies. Many of these companies served various industries, so we could not use official statistics; we had to get closer to them.

We sent out a questionnaire to over 300 companies which we followed up with direct contact, trying to obtain data about their activities. Our focus, to begin with, was mainly on technology, byproduct, and service companies.

We asked the following questions:

- What is your turnover and how much of that is linked to the seafood industry?
- How many employees are directly or indirectly involved with servicing seafood companies?
- Do you collaborate with other companies or institutions related to seafood?
- Do you develop your own technology, or do you mainly import technology/service?

As always the response was very low so we had to visit the companies. The team of students mapped specific areas in Iceland such as Reykjanes peninsula and southern Iceland. These were areas with a population of around twenty thousand and were fairly easy to access, but it meant going door-to-door in many cases. In these mapping projects, we focused on all companies involved in fish processing and byproduct utilisation. This was a valuable exercise and gave us vital data about companies which were geographically close to one another, but culturally distant. Most of them had no formal contacts or relationships with each other; in many cases, they had never even met!

We also did a thorough mapping of ocean-related industries in the Reykjavik area. I wanted hands-on experience for this project, so I visited a number of startups and established companies in Reykjavik. As I write this I still remember the importance of these visits where I had the opportunity to meet and discuss with people from mostly small companies about their activities and plans. In nearly all cases, my visits were very well received and gave me an extremely important knowledge about the activities of these companies. These visits were also important as a first introduction to the cluster ideology and, later, as an invitation to join the cluster.

Many cluster specialists view cluster mapping as a tool to enable systematic comparison across regions. I view the mapping tool as a way to get a thorough profile regarding all major players in a particular area. We need to know about all the players within the region to be able to assess the opportunities they may provide — and their patterns of collaboration.

The team of students who worked on the mapping.

In our cluster work, mapping has become one of the most crucial tools to accelerate collaboration and innovation. Later in this book, we will describe different types of mapping, like mapping fish waste in the North Atlantic seafood clusters, etc. We are also continuously mapping specific fields of industry. Most recently, we have mapped everybody who is into sushi and seaweed in Iceland! We have become mapping nerds and we are proud of it.

In 2011 the cluster initiative published a thorough report on our cluster mapping. This report was edited by Vilhjálmur J. Árnason, former deputy CEO of the Iceland Vessel Owner Association. The report listed our findings as the core parts of the Iceland Ocean Cluster. This mapping was supported by both the in-depth economic analysis and the data collection by contacting various companies directly.

The start of the IOC

The Iceland Ocean Cluster was formally established at an open meeting at the Reykjavik harbor, May 24, 2011. There, we presented the mapping of the IOC. This was a very memorable day, mainly because we were introducing a new and wider lens on an established industry. We showed a picture of a cluster which linked together different parts of the economy — parts which were seldom thought to be intertwined. The media was very positive, and I believe that our detailed and thorough work in data collection and analysis made the difference.

The following picture shows the main parts of the Iceland Ocean Cluster as we displayed it in 2011.

The sub-clusters of the Iceland Ocean Cluster.

As a result of our mapping, we introduced eleven potential sub groups. Each group is represented by a bubble. These groups were:

- Fisheries and fish processing
- Ocean technology
- Transportation and harbors
- Marketing and distribution
- Ocean surveillance and administration
- Finance and service
- Seabed
- Ocean tourism
- R&D, education, and training
- Biotechnology
- Aquaculture

The paradigm shift

The picture below depicts Statistics Iceland's finding that employment in seafood (basic fisheries and fish processing) was diminishing. With this picture, another downward line would indicate that, with new technology, there would be less need for people in the industry.

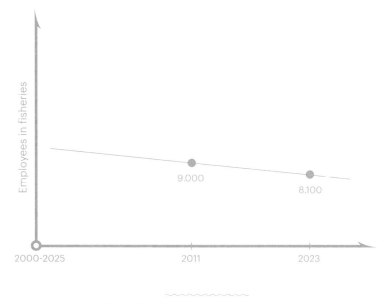

EMPLOYEES IN FISHERIES AND FISH PROCESSING IN ICELAND

"…major parts of the seafood industry are not accounted for as a part of the seafood industry: tech companies, companies in full utilisation of seafood, R&D, etc."

Academics, media, and politicians have talked about the importance of fisheries in Iceland, but many believed that the industry was not providing opportunities. This belief was supported by the national statistics about employment in the traditional seafood industry.

We were proving them wrong. At the founding meeting in Reykjavik, we stressed that major parts of the seafood industry are not accounted for as a part of that industry: tech companies, companies in full utilisation of seafood, R&D, etc.

The picture below showed our take on the whole ocean cluster in Iceland. This gave way for a great and positive story building, based on hard economic as well as qualitative data. We realised we had an opportunity to bring a more optimistic message to Icelanders about the future of this industry.

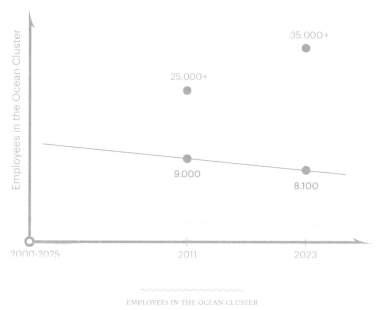

EMPLOYEES IN THE OCEAN CLUSTER

"We realised we had an opportunity to bring a more optimistic message to Icelanders about the future of this amasing industry"

With sound economic data and further mapping of ocean industries, we were ready to go to work. We were able to bring a new message of opportunities in the ocean cluster to the public and to the industry as well.

There was a possible paradigm shift in the pipeline! Our next step would be to establish a leadership group.

Finding the leaders

Leadership in clusters is the art of making the right decisions and then getting the enterprises to want to carry them through. In 2017, I spoke in front of an audience of CEOs and owners of seafood giants at a leading global conference, The Groundfish Forum, on whitefish fisheries. My message — as always — was how to use more! I had even put in some extra slides about possible new revenue streams from byproducts. I felt this could be "the" moment where we would get a powerful group of companies to carry these ideas through.

My speech was very well received and many CEOs showed interest in my message. I found strong interest from leading seafood companies around the world to follow our full utilisation of fish model — 100% Fish!

This conference showed how powerful such an elite industry group can be. However, I believe their mistake was to keep the "club" very tight. I believe the mistake of so many industries is to keep up the club culture.

In most cases, leaders in fisheries have their own association. So do small boaters and processors. At the same time, we know our opportunities lie in combining knowledge and experience. This is especially true in natural resource industries which need continuously to move further up the value line and bring in people from other disciplines to assist. We all know that problems are best solved by sharing ideas over traditional disciplines. The best way to fight information silos is to introduce interdisciplinarity.

At the initiation of the Iceland Ocean Cluster, we were anxious to make sure the leadership group would represent a wide range of knowledge and experience. All members of the board of the Cluster were required to have a signed agreement in which they made a financial commitment to the net-working activities of the cluster. At the same time, we needed to make sure that these companies did not feel as if there were lots of free riders.

Forming the cluster's leadership group was a challenging task. An aspir-ing cluster founder — anyone who is most interested in setting up ocean clusters in their own country or region — needs to be particularly mindful of their leadership team. An ideal team would be comprised of dynamic, engaged individuals who also represent the cluster's diversity.

I had prepared a detailed package for possible candidates for the leadership group which consisted of a draft of the economic analysis, the mapping and examples of success stories from clusters in other regions. As I did not have many examples of ocean clusters, I used other examples to emphasise the opportunities in clustering.

My first meetings were with Guðmundur Kristjánsson, owner and CEO of one the most successful fisheries in Iceland. Even though I knew Guð-mundur from my past work, I was not sure whether he would get my ideas about a value creation in a seafood cluster. I emphasised the ideology behind clusters in the most practical way possible. I told him we wanted to group not only fisheries together but also other parts of the seafood value chain. Guðmundur told me he was especially interested in education and more efficiency on board fishing boats; I was glad to show him that we had focused on education in our mapping.

This meeting was a great morale boost for "a lonely founder". In this first meeting, I realised how important it was to engage with leaders in industry and get them on board the IOC. Like the other introductions which followed, I became more and more aware of their interest in a "hands-on" approach. After my past experience with traditional work for industry associations, I also realised I had to make sure people believed we would make a difference and a bit of a paradigm shift when it came to our work; we were not going to be a group of people who just met to chat about high taxation or regulatory burdens. We would show results in creating new business opportunities through networking.

I went on to talk to more fisheries and their supporting services. The result was a group of ten leading companies which together formed the initial founding members of the Iceland Ocean Cluster. These leaders were Birna Einarsdóttir, CEO of Islandsbanki; Sigurdur Valtýsson, CEO of TM Insurance; Guðmundur Kristjánsson, CEO of Brim Seafood; Gísli Gíslason, CEO of Faxaflói Harbors; Gylfi Sigfússon, CEO of Eimskip; Aðalsteinn Ingólfsson, CEO of Skinney Þinganes; Grimur Sæmundsen, CEO of Blue Lagoon; Árni Oddur Þórðarson, CEO of Marel; Jóhann Jónasson, CEO of 3X technology, and, last but not least, Gunnar Már Sigurfinnsson, CEO of Icelandair Cargo. Later, the following leaders joined our force: Ásbjörn Gíslason, CEO of Samskip; Helgi Anton Eiríksson, CEO of Iceland Seafood International; Gunnþór Ingason, CEO of Síldarvinnslan; Katrín Pétursdóttir, CEO of Lýsi; Helgi Magnússon, Board member of the Federation of Industry; Jóhann Oddgeirsson, CEO of Samhentir; and Haukur Óskarsson, MD, from Mannvit Engineering.

These companies were from fisheries, fish processing, banking, insurance, fish processing technology, and engineering. We emphasised the fact that the members in the leadership group had prior experience in leading companies. Our next move was to get institutions and Universities on board. We were extremely pleased with the Universities' response. Both the University of Iceland and Reykjavik University were interested in close collaboration. We had more difficulty with some Government institutions, which I believe saw us as competitors — even though we did not receive any support from the Government. Later, we built a strong relationship with various specialists in many of the key Government institutions in Iceland.

GUÐMUNDUR KRISTJÁNSSON, ICELANDIC FISHERMAN AND CEO OF BRIM

Guðmundur was the first business executive I introduced the idea of the IOC to.
This meeting was a great morale boost for "a lonely founder".

The leadership has always consisted of around ten leaders who have been with us from the beginning. It is important to keep the dynamism in the group and have a combination of both institutions and business. In our case, the business leaders have always been in majority, but leaders from startups, public institutions and universities have made the leadership group more fruitful and dynamic.

Building trust

In nearly all the regions I have visited, the overall message is always the same. There is very limited collaboration and trust in the industry. There is widespread urgency for more collaboration and breaking silos. Nowhere have I seen a perfect communication between fishermen, processors, R&D, Universities, exporters …

The leadership group within a cluster has to become champions in helping to build trust and conversations across the sector. The leadership group is in a very good position to drive the conversation. They must get other CEOs, from large and small companies, to realise that in this period of transition and competition, "being good at what you have done for years" is not enough.

Somewhat contradicting my message at the beginning of this chapter, I believe forming groups of relatively similar companies is good to start with. The reason is that people in these companies understand right away how they may be linked to people in a similar industry, they are likeminded and they have similar customers and markets to work on. It is a hard sell to a person who is working on basic processing of fish that he or she needs to meet with people at a very different place on the seafood value chain. The groups of similar companies can act as centers of capability. We decided to create the sub-clusters, such as ocean technology, transportation, fisheries, etc. We started to build trust within that group and got them ready to meet with other parts of the value chain in the near future. The cluster manager has to find ways to connect the silos instead of splitting them up.

There is no formula for building trust. However, we can easily increase the likeliness of building trust by bringing people together. The first meetings held by the IOC with the sub-cluster of tech people were very memorable.

It seemed like the techies predicted beforehand how the meeting format would be (they had attended many meetings in the past): You have a cup of coffee, have a seat in a large conference room. You know relatively few and you are not sure whether you have a common goal. You leave the meeting with very little to talk of. Most likely, the only thing which uniting the group would be some kind of message to the government about high taxes or too much regulation.

Our cluster meetings were different: small groups of selected people with clear, practical agendas.

I believe the main thing here is for the leadership group and the cluster manager to tailor-make each sub-cluster. Make sure that the members in the group are interesting to each other. People need to look around a relatively small table and think to themselves, "these representatives are interesting and I believe I may benefit from this meeting". Small groups are a good way to get more people involved.

To begin with, the groups which we formed were basically people from the same part of the industry. It was obvious that many of them knew about the other companies, and right away many saw an opportunity to get to know each other a bit better. Then came another meeting and another. We were sometimes getting distressed that the participants would be asking about the results of the meetings. However, at the 4th and 5th meeting of the tech group everyone seemed to like to meet "their group" (we had the same experience with a group in logistics, fish processing and education). Trust was building up. The other obvious thing was that out of a group of 8–15, often groups of 2–3 CEOs would be gathering after the meetings to chat. This was in some cases more than a chat: they were actually building more trust between themselves and moving to the next level where they decided to work on projects together. Some of these projects have never been defined as cluster projects, but they originated in the cluster. In the logistics group, two companies, from very different angles, met afterward and later established a startup to service cruise liners coming to Iceland.

The conclusion here is that a tailor-made small group is very important. It allows the building of greater trust.

Finding pioneers in byproducts

There is a group of pioneers in byproduct treatment in many fishing towns who are often overlooked. These are the entrepreneurs who are already processing remaining raw materials into basic animal feed.

The animal feed is not a high-value product, but it is an important starting point up the value chain ladder. Many of these companies are owner-operated: they are great innovators who share their ambitions to do more with their products. Many are cash-drained and not strong marketers of their companies to investors. These entrepreneurs, treating valuable proteins from the sea, are often wrongfully depicted as the "waste collectors" in the industry.

We should support and embrace these entrepreneurs. I have met them in fishing towns in Alaska, Norway, New England, France and in other places around the globe. These are people who are often under-represented and under-appreciated, but they show remarkable persistence!

Designing a strategy to decrease waste in the seafood industry is a challenge. The first step is to map all the pioneering companies in "waste collection," map the new startups and to connect the two. The second is to find ways to encourage them to collaborate. One such method has been successfully implemented in Iceland.

Also, realise that the whole group will rarely become a part of a particular project; companies have different interests and trust is not equally divided among group members. Jeff Bezos, the founder of Amazon, suggested if you can't feed a team with two pizzas, the group is too large. Our experience is closer to one large pizza (with at least two toppings!).

One has to stress that however many members are in the leadership group or in the sub-clusters, it's all a question of the willingness of its participants to demonstrate certain flexibility and to communicate and share information. Through cooperation, it is possible to increase innovation; through innovation comes business and production efficiency. In most cases, difficulties are associated with players that have little experience in cooperation and that lack trust. It can take some time to build confidence and open dialogue at a strategic level, and there may be deep-seated animosities between stakeholders (known only to them). Stakeholders may also have different agendas, and they can move at different speeds and react differently. Major companies may feel threatened, and ego clashes between stakeholders can occur. At earlier stages, these obstacles can be more easily overcome by the sub-clusters, where the connections are more obvious and people run into each other frequently at expos, conferences, etc.

We were successful in forming various groups and trust was building. We were collecting strong partners who had the ability to become the main drivers of the cluster activities. However, we knew the next phase was to be our cluster's most important task: to find the "low hanging fruits" and to start the cross-pollination of sub-clusters.

Low hanging fruit

In the early days of the IOC, I realised that if business people were to be active in the cluster, there was a need to keep them busy with interesting projects and low hanging fruits. "Low hanging fruit" is a metaphor used for doing the simplest things first, resulting in delectable payoffs. We all know from endless organisation meetings, whether private or public, that people are generally tired of meetings for the purpose of meeting!

Early on I strove to get groups of companies — often smaller companies — in the industry to meet and start finding the low hanging fruits. These "fruits" were projects which could result in concrete output within 6–12 months. The emphasis was on creating real spin-off projects, joint efforts, and startups. This is a never ending story for a successful cluster and actually the greatest fun as well.

In a cluster manager's dream world, new projects continuously fall from the skies: people can't wait to meet and talk about ideas, the trust level is high, and things move fast. This has seldom been the case for the IOC, even though we believe Iceland is a country with a fairly high trust level and openness. Clusters need a dedicated workforce that is motivated to show results.

The mapping and economic analysis we had done gave us the directory to all major parts of the cluster and made it easier for us to start fitting together various companies — at least in our minds. We knew, however, that we needed to sell the idea of collaboration by emphasising the low hanging fruits.

In this chapter, I will describe some of the initial projects we undertook and how they have fared.

Green Marine Technology

Low hanging fruit ▽ Marketing green technology
Goal ▽ Strengthen the image of Icelandic green technology
Participating ▽ Tech companies and artists

Iceland is surrounded by some of the richest and most prolific fishing grounds in the North Atlantic Ocean, and fisheries have long been the mainstay of the Icelandic economy. Healthier oceans and environment are, therefore, a matter of fundamental importance for the country.

In the interviews we did in the mapping stage, we saw a strong "green" trend among seafood tech companies. However, the green features of their technology were not heavily emphasised. It may be partly explained by the fact that green technology comes quite naturally in a country with clean energy sources. Though most of their technology was based on hydro- or geothermal power, there was a clear opportunity to market these solutions under a more green label.

~ Is startup the key? ~

Even though academia and business may be genuinely interested in working together, the execution of good ideas is always a challenge. This is a puzzle for many ocean enthusiasts. Even though business may be very interested in new projects, the company's staff is often kept busy with their day-to-day tasks. It can be hard to allocate company resources for exploring new ideas.

I believe here we need to include the startup world and risk capital. It means that a special project (which of course needs to be well defined and have independent cash flow opportunities) is made into a limited company, a spin-off. For it to qualify as such, the project must get a well defined budget and management. Also, for many existing seafood companies, this is allowing them to keep doing what they do best at the same time as they "work on" new projects which do not interfere with their core business.

This also opens doors for venture capital funds to join the seafood startups early on. Including risk capital from venture funds at early stages can also calm the seafood companies, who may worry that most of the project financing would otherwise be in their hands.

The idea was for the companies to introduce themselves as providers of outstanding green technology and contribute to the improvement of a better environment. The technological solutions are based on better utilisation of energy resources, reduced oil consumption, better utilisation of raw material processing, and more.

We were really pleased to to find that the tech companies were interested in this possible collaboration project. All the members of the project had to put a relatively small amount of financial resources into it so that we could start the project.

Even though the project sounded simple, it took 18 months for ten Icelandic tech firms in the fishing industry to sign a cooperation agreement for the development of green technology in fishing and processing. The project was named Green Marine Technology. The President of Iceland, Mr. Ólafur Ragnar Grímsson, launched the project at our headquarters.

The aim of the project was to strengthen collaboration among technology at the leading edge of Icelandic technology. At the same time, the project drew attention to the Icelandic fisheries' global leadership role in high-quality fishing and fish processing. The group decided that the web site would be ready in 6 months and signed up a team of designers to make an artistic expression of the companies' technology.

At later stages, we realised that ten companies in a group is a challenge. We often find two or three companies taking the lead and another two or three are simply along for the ride. In between are companies which are interested in the meetings but are not very active. This is one of the difficulties of driving projects which are not financed properly from the beginning. Both of these problems surfaced during this project: because the project was not well financed, companies were putting different levels of energy and time into it. However, what we found is that offshoots from the Green Marine Technology group were occurring. Often, two or three companies within the group started collaborating on specific tasks relating to concepts like green marine, green fishing vessels and green fishing gear.

I am very pleased to see how companies and startups in the ocean cluster in Iceland are emphasising saving the environment by introducing better and greener technology. We see different types of collaboration, at least some which can be traced back to the cluster work in this field. Just to name a few interesting green projects:

- New bottom trawl doors that do not touch the seabed have been invented. These are controllable doors that float above the seabed, reducing seabed contact and drastically reducing energy consumption.

- IT companies are focusing on solutions that can increase efficiency and traceability, saving energy and reducing pollution.

- With new fish processing technology comes more yield and higher seafood quality — hopefully, we can do more with less in the fishing industry with these technologies.

- Engineering firms in the field of naval design are introducing energy-saving ships: electric boats with no CO_2.

- Green nanotechnology is used to power fishing boats.

- New cooling technology on board ships does not need ice which saves energy and increases quality.

- Electric winches on board ships yield the same results as traditional hydraulic winches but use much less energy.

- Cleaning and disinfecting technology for fish processing plants use only environmentally safe material.

All fishing nations have learned that they need to be responsible when using their resources. Environmentally friendly technology can improve our lives in coastal communities, increase efficiency, and reduce waste; threfore, it is good business. However, in my mind, there is little or no focus on green technology on board the global fishing fleet. We have amazing inventors and startups in many countries who have developed green technology like electric fishing boats and environmentally friendly fishing gear. Unfortunately, many of those inventors are not receiving the support they need. Moving forward, Green Marine will continue to be an ongoing task for the Iceland Ocean Cluster.

KÁRI LOGASON & SVERRIR BJARNASON

Kári, a naval architect, and Sverrir, a megatronics engineer, are a part of the new generation of naval architects in The Ocean Cluster House in Iceland who are at the forefront of environmentally friendly ship design.

The Green Fishing Vessel

Low hanging fruit ▽ Prototype for a green fishing vessel within 18 months
Goal ▽ Develop an electric longliner
Participating ▽ Naval architects, tech companies, fisheries, and Government Research Fund

One of the projects which resulted from the Green Marine Technology group was the Green Fishing Vessel project. Six tech companies and one fishery met in 2013 to discuss a possible collaboration on building an electric fishing ship. The group was interested in searching for ways to increase the opportunities for fisheries (long liners) to utilise electricity on board the ships to lower fuel consumption and emissions, which is a major issue for ships and fishing vessels as it is estimated that fisheries consume 40 billion liters of fuel and generate a total of 179 million tons of greenhouse gasses (Parker, R. 2018). At the initial meetings, attendees were skeptical that there was not enough interest in the industry for making an electric fishing vessel, since it would be considerably more expensive than a traditional fishing vessel than traditional fishing vessel. Government support in some form was needed so the boat could become a reality.

For the next three years, various collaborative efforts worked to bring Government or Government funds to the table. This was also the first project where we explored collaboration with companies in Norway. The group wanted to investigate alternative designs for the propulsion system of a fishing vessel, including the probable effects related to operational conditions and model accuracy with a focus on the impact on fuel consumption.

We were fortunate to start a collaboration with The Maritime Cluster in Norway. The clusters aimed to share knowledge and explore possible collaboration in product development between member companies regarding the development of fishing vessels. The strength of Norwegian ship building lies partly in Norway's extensive experience with design and construction of various specialty vessels. This has resulted in the development of large Norwegian firms which concentrate on marine design and construction. Iceland's strength lies mainly in the highly demanding fishing industry, which has required ambitious advances in technology for use on board fishing ships. However, the lack of variety in ship demand in the country means that marine tech and engineering firms are very small and have difficulty expanding to other markets. The close collaboration between tech firms, engineering, and fisheries has often resulted in efficient technology. However, tech and engineering firms suffer from a lack of market connections and limited size.

We set up a working committee to connect Norwegian (and other NCE Maritime member firms) to Icelandic fishing vessel tech and engineering firms to enhance relationships, build trust, and hopefully develop mutual products. The first meeting of the working group was held in Ålesund, Norway, November 2013. Eleven representatives from nine technology companies from Iceland attended the meeting and met with representatives from various Norwegian companies. The project "The Electric Fishing Longliner" is now led by one naval architecture firm in the cluster, NAVIS. It has received public grants, and we expect the first electric vessel to be built in Iceland before the year 2022.

Ocean Excellence

Low hanging fruit ▽ Develop a team of specialists for innovative processing solutions
Goal ▽ Become leaders in small scale solutions for artisan fisheries
Participating ▽ 3 leading enterprises

At one of the earliest meetings of the Iceland Ocean Cluster — with a variety of technology, engineering, and fisheries enterprises present — the group wanted to address challenges facing developing nations in processing fish and ensuring its shelf life. In many cases, the solutions that Iceland had introduced were not suitable for many developing countries or very small fishing villages. The Icelandic solutions were made for large-scale production where electricity needs were easily met; bringing these solutions to smaller places was impossible. Instead, the group discussed possible solutions regarding either smaller-scale systems for drying fish or cooling fish. The start-up company Ocean Excellence (OE) was formed by three major enterprises in Iceland in collaboration with the Iceland Ocean Cluster. The partners included Mannvit Engineering, one of the leading engineering companies in Iceland; Samey, a leading company within food-grade automation and artificial intelligence solutions; and Haustak, a cutting-edge company focused on full utilisation of Iceland's white fish resources.

The initial matchmaking was a success: the team proved to be a great mixture of players from different fields of expertise to support the development of new and creative solutions. During its business, OE has promoted solutions within the turning waste to value concept. OE has also worked with US based aid NGO's to develop sustainable storage solutions to be applied within artisan fishing sectors.

FISHING SHIP OF TOMORROW

The "Fishing Ship of Tomorrow" is displayed in The Ocean Cluster House and presents various Icelandic innovations on board the modern fishing ship.

The birth of Codland

Low hanging fruit ▽ Find ideas to increase cooperation in more use of fish discards among "seafood" enterprises within 12 months
Goal ▽ Utilise more of each cod
Participating ▽ Fisheries, Engineering companies, tech companies, skin care companies and government research fund

The greatest challenge for a seafood cluster manager is to get together a group of fishermen. They're not the greatest fans of meetings! In Iceland, some CEOs of fisheries are very powerful in their communities: many have been driving forces behind various social impact projects and are interested in strengthening their local community. These folks are not anxious to attend more meetings. They have their own association which deals with wage negotiations, relationships with the fisheries ministry, lobbying, etc.; lots of issues, many unsolved.

"What is a cluster? Will it do anything for my community?". I had to prepare to answer these questions and many like them. What had impressed me the most in the mapping phase was meeting various entrepreneurs and researchers in byproduct utilisation. I thought it would be wise to offer the fisheries the opportunity to meet the new startups and get a glimpse of the "new seafood industry". Our "low hanging fruit" task became "finding ways to increase cooperation and use of fish discards among seafood enterprises within 12 months". We decided to have our first meeting in Grindavik, one of the most dynamic fishing communities in Iceland. We planned presentations from companies which were developing consumer goods like skincare products and skin supplements from fish.

To make a long story short the meeting was a great success. The fisheries CEOs from the community were anxious to know more. I will always remember one of them saying to me right after the meeting:

> *I did not know someone was making skin care products from the intestines of my cod.*

The skin care company had already been operating for more than 20 years, but mostly within the University system. After our meeting, the CEO of the fishery and the skin care company planned a get-together at the CEO's office; the skin care team told me afterward that the meeting had been very successful. The skin care entrepreneurs told me this was the first time they had entered an office of a CEO of fisheries.

The "Grindavik meeting" can be identified as the game-changer for the Cluster in terms of starting to make the connection between startups and fisheries. These relationships have since become a very fruitful part of the Iceland Ocean Cluster work. The meeting also laid the groundwork for our accelerator function in full utilisation of seafood.

Codland started as a cluster project which emphasised Iceland's ability to become known worldwide for the sensible and sustainable management and full utilisation of the Icelandic cod. We argued (humbly) that Codland would become the Silicon Valley of cod! In some of my earliest writings about Codland (2012) I was pompous enough to start citing Silicon Valley as our prototype:

> *The electronics hub in Silicon Valley, California has become a world leader in electronic development. Many of the greatest innovators in IT technology were initiated in the valley such as HP, Apple, and Intel. This development did not happen overnight, but there were certainly some important parts which existed in the Valley which made it all work. What of Silicon Valley's example can be used in the "fish" context? Silicon Valley is built on a certain mindset: belief in the individual's abilities to create strong cooperation between superb universities and industry, a tradition in electronics manufacturing, and a community of practice — people share ideas and risk.*

And urging for more action:

> *There is still a lot of work to do in Iceland to live up to its name as a leader in cod in the world. Countries which enjoy*

PÉTUR PÁLSSON, CEO OF VÍSIR FISHERIES
*Pétur played a crucial role in
developing Codland.*

TÓMAS EIRÍKSSON CEO OF CODLAND
Tómas's ability to build further bridges between R&D and industry was a very important part of the success of Codland.

rich natural resources often suffer from resource blindness; they don't see the opportunities in their own resources. This is still somewhat evident in Iceland. First, we have not succeeded in encouraging young people to pursue careers in sustainable fisheries, seafood biotechnology, and fisheries management. This is worrisome. There is no future COD-LAND without well educated and enthusiastic new generations. Second, even though R&D has increased in the seafood sector in Iceland there is much more needed to develop further products and know-how in fully utilising the cod and other seafood products. More investment in research, innovation and technology is needed.

Codland was a good example of a project that could have ended up as a great idea and beautiful intention but lacking output. I found out early that it was difficult to find ways for a large group of fisheries and byproduct companies to collaborate on a non-profit level. Companies were just not used to it, and the "free rider" problem seemed to be evident early on in the process. So my next steps were to establish Codland as a limited company and bring key companies into it early on, with serious intentions and real financial commitment. This changed the pace and somewhat the culture of Codland. Today, Codland has become a bit of a household name in Iceland with strong leadership and a clear mission to use the fish 100%.

With Codland came our strategy to be a more active accelerator and investor than originally planned. This move strengthened the cluster financially. When we founded Codland we saw that there was a gap between R&D and industry. The gap meant that new startups were not being created and brand new product ideas — especially regarding various byproduct utilisation — were not activated. With Codland we knew we had an important role to play. This is the first project where the IOC established a limited company (Codland Ltd.), owned the equity, and later sold the majority to fisheries. A special venture capital fund, owned by the IOC, has since become a stakeholder in various startups. The fund has equity in startups, either through seed investment or as an entitlement for being one of the co-founders of a company.

The importance of risk capital in an ecosystem is often overlooked — and so is the role of risk capital in clusters.

> *In Western innovation policies, private risk capital was largely irrelevant. That is no longer the case. Looking to innovation hotspots like San Francisco, Tel Aviv, Beijing and Stockholm, private, competent risk capital matters. Today, you cannot build a strong ecosystem without a large number of capital actors (hint: this is not your local credit- and savings bank). With growing access to capital, good ecosystems are able tap into, raise, and activate larger amounts from highly capable investors. A dynamic and liquid ecosystem will also create opportunities, investments, rapid investment rounds and — hopefully — strong exits. This again fuels the next generation in the ecosystem, creating a positive reinforcing mechanism, allowing strong ecosystems to get stronger. (Rangen and Food-Hodne, 2019).*

Marine Collagen

Low hanging fruit ▽ Make a feasibility study regarding fish collagen
Goal ▽ Build a plant in Iceland with a capacity of 3–4000 mt of fish skin
Participating ▽ Fisheries, government research fund, engineering company

Collagen has been extracted from fish skins for at least 50 years. Still, the commercial volumes of fish collagen have been relatively small compared to other animal collagen. Was there an opportunity? In 2012, I came across an article about a marine collagen plant in the Faroe Islands which had gone bankrupt. The plan had been to process fish skin into pure marine collagen protein. I soon came to know that a Spanish company, Junca Gelatin, had been involved with this project. I contacted the company's leadership and realised that they were very positive towards cooperating with us on further development of a marine collagen plant in Iceland. The initial idea was to buy the existing equipment from the Faroe Islands and set it up in Iceland.

HRÖNN MAGNÚSDÓTTIR AND KRISTÍN ÝR PÉTURSDÓTTIR, FOUNDERS OF FEEL ICELAND

Are women taking the lead in the high value end of the seafood industry?

At this point, we had a team of students from the University of Iceland who were collecting information about marine collagen and doing a simple feasibility study regarding a possible plant operation in Iceland. The first study indicated Iceland had a sufficient amount of fish skin available to efficiently process marine collagen and world market prices were quite favourable.

In 2013, we brought in two large companies from the cluster network to lead the cluster in developing this idea further: fishing company Vísir and Mannvit Engineering. We created a thorough plan for a new manufacturing facility in the Reykjanes peninsula with the aim to manufacture collagen from cod, haddock and saithe skins. The process was based on washing, pretreatment, extraction, filtration, concentration, sterilisation and finally, drying.

The treatment of skins and the technology behind the factory were known. However, there was limited information available regarding the characterisation of collagens from different fish types. We sought grants from government research funds to map the quality of different types of fish skins and to increase standardisation in the process.

In one of the memorandums we stated:

> *The new facility will manufacture collagen from fish skins, which in many instances are currently discarded or exported for further processing to other countries from Iceland. The project, therefore, has the potential to create value for the Icelandic economy and increase knowledge in full utilisation of the fish.*

This company was a typical project which was initiated by the cluster. Later, more fisheries joined the Marine Collagen team. All these fisheries have been directly involved with the Iceland Ocean Cluster.

In 2019, the collagen plant was formally started and this company is becoming one of the cluster's pride: a great part of the "new seafood" of Iceland.

Feel Iceland

Low hanging fruit ▽ Find a team to market marine collagen for the consumer market
Goal ▽ Strengthen knowledge in retail marketing of final products
Participating ▽ Entrepreneurs with business/marketing and design background, investors, R&D Funds

When we were working on the collagen business plans, we knew we needed to accumulate more knowledge and skills in retail marketing and sales of health products, such as pure protein or skin care products. We needed a marketing team.

In 2013, three women entrepreneurs expressed interest in joining us. All of them were somehow connected to fisheries and they had an interest in marketing high-end fish fillets. We suggested this team should observe opportunities in marine collagen.

To make a long story short, this team took flight, created Feel Iceland, and built up a very successful company. The Feel Iceland team started early to strengthen their network by partnering with many companies and investors within the IOC. These partners included fisheries which could provide the fish skin, startup funds which had already been backing many interesting projects within the IOC, and product designers.

Early on, we noticed that even though the Feel Iceland team did not have a background in chemistry, their business, marketing, and design skills meant that they were quick to use the IOC "kitchen" — filled with ingredients and utensils for a successful startup. They were super networkers with a great story to tell. I was very pleased to see how the IOC was a great platform for Feel Iceland. They have received grants from public funds and new shareholders have invested in them.

Bringing small fisheries together

Low hanging fruit ▽ Strengthen marketing and sales of mackerel within 6 months
Goal ▽ Build a collaboration among small fisheries resulting in better long term fish prices
Participating ▽ 10 small fishing companies

One of the challenges of The Iceland Ocean Cluster has been to inspire small fisheries to collaborate. The most interesting project of that kind was initiated by Kristinn Hjálmarsson, an entrepreneur who had the ambition to get mackerel fishermen, who catch with hook and line only, to collaborate regarding marketing and sales of mackerel. This method of fishing promotes two fundamental elements at the same time. On one hand, the catch is clearly sustainable and ecologically friendly. On the other, the catch is superior in quality as each fish is individually caught, bled, cooled, processed and packed.

Mackerel closer to the shore are larger, and because of that, small boat fisheries are catching the most valuable mackerel, as they are outside the reach of bigger vessels. It is a good example of how different fishing methods can compliment one another — as long as all policy makers, in different countries or unions, are being responsible in their decisions on a total allowable catch from every species.

The cooperation among the local small boat fisheries could improve raw material prices through higher levels of quality. Increased quality begins on board the small vessel, where cooling and handling are crucial. It is impossible to market and sell a high quality product as such, if it is not handled with care and landed in prime condition.

Mackerel from small fisheries passes the three-tier test of sustainability with flying colors. Catching mackerel with a hook is ecologically friendly: it has no effect on the ocean's ground floor and other species do not bite on the mackerel hook. Catching mackerel with a hook makes great economic and social sense, as the approach brings life and action to small and large communities which may depend largely on fisheries. In this day and age,

KRISTINN HJÁLMARSSON
Kristinn led a project to bring mackerel fishermen to collaborate.

the small fishing village may not see as much activity and employment as before.

The collaboration led to higher prices for the inshore mackerel fishermen and was, at a time of strong market conditions, an excellent example of cluster collaboration. All the small fisheries contributed financially to the project, but the management of the project also had incentives to sell at higher prices. Early on we saw this project as a prototype of other projects which could follow: collaboration among shellfish fisheries, longliners in whitefish, etc.

As market conditions for mackerel got sour, the situation became quite different. One fishery received a few percentage points higher offer for their mackerel so they suddenly left the co-op, which meant that the morale in the group changed. Suddenly, the many small fisheries were not as interested in collaboration as before.

The weakness with a large group of small fisheries is definitely their lack of high-trust relationships. Many of the small fisheries have had their issues with each other in the past and there has been competition.

This is very similar to what I have since learned to be the case with many collaboration projects among small fisheries in other coastal areas. There is lack of trust, especially if the group exceeds four to five small enterprises. There is no need to give up on such projects, but it takes time to build lasting relationships among small fisheries.

The Logistics Sub-cluster

Low hanging fruit ▽ Get a dialogue going between all parts of logistics in Iceland
Goal ▽ An industry perspective on logistics strategy for Iceland
Participating ▽ Logistics companies, banks, harbors, airlines, engineering

The logistics sub-cluster of the IOC was established in 2013. This sub-cluster was prototyped on the lines of global logistic clusters. We noticed that logistics clusters consist of companies representing all parts of the logistics chain. We realised that such a network did not exist in Iceland. The harbors have their own association, airports are a national government organisation, airlines are within the Employers Association, etc. It was clear that conversation between different parts of the logistics industry did not occur.

We started a conversation between the companies in the logistics industry. To make the meetings more interesting we asked CEOs of different logistic companies to host separate meetings. It seemed CEOs were quite pleased to visit each other and learn more about their activities.

We were very careful not to break any competition rules by bringing together competing companies at our meetings, and we received advice from one of Iceland's largest law firms on this issue. In the early days of the logistics meetings, we were observing opportunities to create similar low hanging fruits with the logistics group as we had with other groups. We were fortunate that one such startup emerged that focused on selling Icelandic products to cruise liners.

At the same time though, we felt the group had more need for discussing logistics strategy for Iceland. This was not in line with our cluster methodology, but we also knew we had to be practical and agile in the cluster. We initiated a brainstorming among the participants group in 2013. This led to a report in 2014 called "Logistics policy for 2030". The three major strategic agendas which emerged from the group in this report were; 1. Logistics and service hub for Greenland, 2. Promoting R&D and education, and 3. Iceland as a logistics and service hub in the Arctic.

The logistics sub-cluster, called "Harbor and transport group", has been very active since its initiation. Yearly, the group holds the only logistics conference of its kind in Iceland where all major logistics companies are invited.

BERTA DANÍELSDÓTTIR
Berta, CEO of The IOC, has successfully managed many cluster projects.
She led the formation and opening of the first Food Halls in Iceland.

The Food hall concept

Low hanging fruit ▽ Can we create a coworking space for foodies?
Goal ▽ Set up a street food hall
Participating ▽ Foodies

Iceland is a food production country. Sometimes I have felt we Icelanders do not understand it ourselves. Food production, whether in fisheries or agriculture, has always been a part of our culture, but we have often treated these businesses lightly. They have just always been there!

The game-changer for us was probably the fast-growing tourism in the last 10–15 years. With the flood of tourists to Iceland in the last 10 years, we are suddenly seeing Iceland through the eyes of our foreign guests. They love our pure food and constantly post photos of regular Icelandic dishes on social media. Whether lightly fried cod or smoked lamb — our everyday life was more interesting than we thought! They have also made us more aware that much of our food is less processed than in many other countries. We use sustainable energy for our food production and sustainable methods for our fishing. I am absolutely sure we did not fully realise this until our guests told us.

These trends affect our work in the cluster. We received many entrepreneurs who were anxious to start their own business in catering and in developing new Icelandic dishes. To begin with, our low hanging fruit was "Can we create a coworking space for foodies?". A food hall offering space for entrepreneurs could be a valuable coworking space for foodies! I had been a fan of food halls in Europe, especially those in Copenhagen, Amsterdam and Barcelona. This dream had been in my mind since I visited the food halls in Barcelona in the '90s.

In 2012 we brought together food enthusiasts who had been working in the cluster. We sent a letter to the Reykjavik Municipality in which we offered to set up "The Reykjavík Seafood Hall" in a space by the harbor which had become available. We created a thorough business plan and teamed up with a group of food entrepreneurs and food enthusiasts in the cluster. We were not chosen for this space but we continued working on the plan, now with the ground floor of The Ocean Cluster House in mind. This was a space which previously had been an old fish factory. We received a positive commitment from the board of the harbor authorities which owned the space, so we started serious marketing and business planning work.

As in all our work, we spoke openly about the idea of a food hall. As we were starting the reconstruction of the old fish factory to become a food hall, the Reykjavik municipality advertised their intent to have an open bidding process for a run-down bus station in the heart of Reykjavik to become a food market or a food hall. We were, of course, shocked. We felt that because we had been openly talking about our ideas, the Reykjavik municipality had taken the idea further. So we decided to bid for that space as well. We got it! To make a long story short, we opened two food halls with less than a one year interval and both of them have become quite successful. It opened the doors for many food entrepreneurs to try out their ideas in a street food-style halls. The cluster also operates a pop-up place in Grandi Mathöll where we introduce new food entrepreneurs. We believe the street food hall culture, which did not exist prior to the opening of our Food Halls, has been a game-changer for the restaurant culture in Iceland. We hope it will lead to further changes, opening more doors for small caterers to cluster in small groups at new food halls to try out new ideas.

Finding more entrepreneurs

Low hanging fruit ▽ Find seafood entrepreneurs faster
Goal ▽ Strengthen collaboration with clever startup accelerators and get more startups going
Participating ▽ Icelandic Startups, The agricultural cluster and Iceland's Culinary Treasures.

With our limited number of staff, we have always sought collaboration with other organisations — and mainly those who are better than we are in interesting fields. One is an organisation called Icelandic Startups. This organisation has an amazing team of people who are helping startup founders to accelerate their businesses at the earliest stages, and connecting them with industry experts, investors and leading startup hubs abroad. The IOC

initiated a business accelerator program for startups in seafood and agriculture in 2018 in close collaboration with Icelandic Startups. This accelerator became a huge success and will be held annually. The idea behind this new annual accelerator is to further connect the IOC with entrepreneurs all over Iceland and use a clever team of startups specialists to mentor the startups in the earliest stages of their development. Through this new accelerator, we found new and exciting entrepreneurs in our field. One of the startups is developing technologies that allow scientists and researchers to navigate efficiently through DNA data and drive the development of new diagnostics and ways to stop diseases. Another startup is developing healthy snacks including cool packaging and branding of dried fish. The sole purpose of this startup is to spread healthy snacking habits. The third startup to mention is a company focused on traceability with a new generation of software services based on electronically certified registrations for food traceability.

And all the ideas which failed

We rarely discuss our failures with our "low hanging fruits" strategy. I promise you, they are many. Many fruits which may have seemed low hanging were absolutely not or they are not as sweet as you thought. Sometimes, cluster managers need to try out things. Just as Ifor Ffowcs-Williams had stressed so often, cluster managers need to have a broad portfolio of projects in the pipeline.

Cluster managers are supposed to seize opportunities, that is definitely the fun part of the job. The truth is, the cluster manager cannot be the project manager for all these projects. As a cluster manager, I have been personally very involved with several projects — breaking one of many rules — and often found myself becoming the project's manager. The reason has been either a lack of financial resources or no leader found for a particular project. So I know from first hand how projects fail. Of course, I blame myself for some of the failures and have tried to learn from them.

I believe that roughly half of the projects we have initiated through the years have failed or have never created any significant value for the members. The difficulty is, it is very hard to find one explanation for the failures. I think in all the cases, the ideas were pretty good. The execution is the problem. You may find enthusiasm at the first meetings, but then the members are busy with the "daily work" and start to lose interest. The group may also lack enthusiasm as time goes by if things do not move as quickly as expected. Also, people in the group are putting different focus and time in the project, which irritates those who are putting the largest effort into it. The project manager may have become a bit too anxious and excited about a new idea, finding out later that the idea is beautiful, but in no way executable in this part of the world or region.

I think the best example was an idea which I was enthusiastically trying to get through regarding a particular type of seaweed. I learned a few months later that there were natural conditions in Iceland which made it very hard to grow in the cold ocean around the island. If I would have just have googled it!

Risk and rewards

To summarise, we have found the low hanging fruit strategy to be a very important tool for the cluster and it has resulted in various successful projects. The projects mentioned in this chapter fit into many of the categories often associated with cluster activities. Our greatest success has been with joint ventures such as such as Codland, Collagen, and Food Halls, to name a few. Many of the IOC projects are focused more on general collaboration with conferences and strategy such as the yearly conferences of the "harbor and transport group" and marketing strategies such as the Green Marine Tech group. I believe the success of the joint ventures lies in their clear focus and funding. Participants have dedicated resources to the project and they know their share in the risk and rewards.

As previously stated, many projects fail or are very difficult to manage because not all participants are putting the same resources and time into them. In some cases we have seen how the "free rider" problem kicks in.
This often ruins morale. In my experience, the optimal size of companies in a non-profit project is 3–4 companies. If there is a history of collaboration between the companies in question, a larger group can collaborate efficiently.

Through competitive bidding, governments can encourage collaboration, especially in projects where the rewards or the benefactors of the rewards are hard to define--like where a cooperative effort may benefit multiple stakeholders. Many projects of this sort will, of course, be found in cluster work. They are very much in line with Goal 17 of the United Nations Sustainable Development program focusing on Partnership for the common good.

The Ocean Cluster House

As previously mentioned in this book, the network relationships in the seafood industry are smaller than in many other industries. Could a cluster facility make it easier for startups and existing small tech companies in seafood to build their network and collaborate? Could a facility make it easier for startups in seafood to access capital, mentoring, and people in the industry? Could an Ocean Cluster House with a mixture of companies linked to seafood prevent the premature death of many startups? These questions were too important for us not try it out!

We opened our doors in 2012.

The Ocean Cluster House is really a unique place. As the home of the IOC, we are at the forefront of entrepreneurial development in the established and emerging marine industries. Here, we network and integrate a collection of highly innovative companies that embody the very best of the core values we represent: innovation, entrepreneurship, and sustainability.

Behind us lies a long and proud history of development in the seafood industries and under this roof, we are taking that history forward. Our mission is simply to use 100% of every fish caught, taking fish beyond the plate, and making it into cosmetics, biotech, and pharmaceuticals. Many of these are young companies with new ideas, and each one contributes an important piece into our puzzle.

The initial idea of an ocean cluster house is older than the cluster idea. In 1993 I was just starting my career, working as an advisor to the Minister of Finance in Iceland. My office had a view over the harbor where a large, rundown warehouse was situated (where Harpa Concert Hall is now). I had been editing a magazine of a student organisation. One of my pieces was an article about a new technology park at the university where small tech startups, often spin-offs from university research, were located. In those days Icelandic tech companies in seafood processing were starting to export their tech products, and I thought a tech park focused on seafood would fit perfectly into a facility like this.

The Ocean Cluster House in Reykjavik opened its doors in 2012.
In the beginning, twelve companies had offices in the facilities, but now there are over 70.

A letter to the Harbor authorities in Reykjavik followed where the idea was introduced. It was well received and I was commissioned to write a report to further the idea. The report came out a year later, in 1992. I was told a government committee had reviewed the report, but it never took flight in the hands of Government.

Once we had established the Iceland Ocean Cluster, it was time to brush off the dust from the tech park idea. Just a few months after the establishment of the Cluster, we brought together a group of entrepreneurs within the IOC who were interested in having a mutual working space. We started a dialogue with the Reykjavik harbor regarding a facility for the Ocean Cluster. We had been observing facilities down by the harbor and one stood out as a fascinating opportunity: an old fish factory and warehouse in the Grandi harbor district. Boats lie moored to the piers behind it, and across the street was a row of old bait shacks. The 2500 square meters had been mostly vacant for over ten years. It seemed a perfect match.

Cross pollination is the key to success of any cluster.
Group photo of the tenants at The Ocean Cluster House.

A group of five tech companies joined me for the first meeting with the Reykjavik harbor. We were received by the CEO of the Reykjavik harbor, Gísli Gíslason, who became a strong supporter of our efforts. It took 16 months to finalise the first phase of the building. For me, it was absolutely great to have my wife, an architect, by my side in the designing phase. An article about The Ocean Cluster House, published four years later, in an HA Design Magazine (Gunnarsson, 2015) says it all:

> *The entire length of the building is divided into offices and meeting rooms with glass partitions. Every minor detail speaks to the artistic arrangement, with every table and chair a designer piece, most of them Icelandic in origin. It is instantly clear that design is important to every aspect of what is done here.*

The Ocean Cluster House opened its doors in 2012. In the beginning, twelve companies had offices in the facilities: there are now around 70 at the time of writing. When the first phase of the facility was ready, we had a grand opening. Many prominent people from the seafood industry joined us and there was optimism in the air, but not everybody saw the potential. One CEO from the seafood industry approached me and congratulated us for this project. He was very happy with the initiative but also commented:

> *"I think I know the industry fairly well and there is probably no need for more space at this time for the companies in the industry".*

Six years later, the number of companies had grown fivefold and there continues to be a waiting list.

The main reason for the success of The Ocean Cluster House is cross-pollination. A recent study by the IOC shows that over 70% of the companies in The Ocean Cluster House have collaborated with another company in this facility. The Ocean Cluster House is self-sustaining and not subsidised by the government.

The Ocean Cluster House is an example of a project which is actually drawn from the cluster manager himself. Since the early days, I have witnessed so many cases where my staff and I are actually in a great position to see the big picture and activate the network; we are meeting different actors and we see right away the strength in bringing them together to push new ideas further. These ideas are often hard for one player to act upon, but easy for the cluster manager who has many entrepreneurs in the network and can make the connection. The Ocean Cluster House has been an excellent base for such activities.

As the Iceland Ocean Cluster is a pioneer in the field of seafood coworking space, I think it can be valuable for others to learn more about our strategy and experience in this field. There are five major elements which have made the Cluster Houses unique.

1   The cluster house is focused on specific groups of industry, but also emphasises the importance of bringing in new services and entrepreneurs who can strengthen the existing industry. For instance, in The Ocean Cluster House, we combine core seafood industry people with startups in IT, product design, social media marketing, etc.

2   The Cluster has its meetings and gatherings in this space. There, all the members of the cluster meet with startups and get a great insight into the dynamism in the startup field. This develops an interesting community which all parties can benefit from.

3   The space is used by startups and larger companies, giving both groups in the cluster a valuable networking opportunity. Here, it is also important to have a mixture of people of different age and gender. We have seen amazing spin-offs where veteran entrepreneurs have collaborated with young entrepreneurs in the Ocean Cluster.

4   The focus on startups and the startup community gives the cluster house a unique sense of this important part of the cluster. Startups are not often paying members of clusters and are therefore often left out. The cluster house opens the doors for these valuable players in the cluster. The Ocean Cluster House has received grants from large companies to

operate a special startup space within The Ocean Cluster House. This means that many brand-new startups can have workspace at a much lower cost.

5 I am often asked what service we provide to startups. Putting aside some absolute basics such as the internet, copying machine and coffee (which is, of course, provided), we want the startups to show a certain character before we start plugging them into our network. If the founders show dynamism, organisation and dedication, we become quite excited to assist them, find investors, invest ourselves, assist them in finding the right contacts, etc. Our strategy is, therefore: if you show dedication, we are all in!

The startup community

It's essential to work closely with the startup community and support its growth. This startup community is very strong in Iceland, and the IOC is able to actively support startup events hosted by various industry associations, universities, and private entities. Our role has been to inspire more entrepreneurs to establish startups in ocean-related industries. As soon as these startups have gone through the initial startup process and competitions, we are ready to nurture them further — offering a close community, assistance with business planning and strategy, workspace, networking opportunities in our field, connections with investors in ocean businesses, and beyond.

We have been quite successful in inspiring and supporting startups in our field. The business value of startups in The Ocean Cluster House in Iceland, which have been in our facilities for the last three years, is approximately USD 100 million.

I believe The Ocean Cluster House has become a game-changer. We changed the simply inaccurate view that the fishing industry's enduring image is overwhelmingly masculine. The Ocean Cluster House is filled with startups that were started by a fairly equal amount of men and women. In 2018, the startups had equal number of men and women as founders. These companies develop everything from environmentally friendly fish pallets to IT for seafood traceability. We want to present the seafood industry as a dynamic force which fosters innovation and startups (which, in many regions around the world, has fallen by the wayside). There is a lot of untapped potential out there. Nowhere is this more evident than in the fact that there are dozens of great startups in The Ocean Cluster House that no one has heard of yet.

The Cluster House as a kitchen ...

I have often tried to explain why a physical space has played such a significant role in the success of the Iceland Ocean Cluster and I believe it is best done by introducing Sarasvathy's concept of causation and effectuation (2001).

> *A simple example should help clarify and distinguish between the two types of processes (causation and effectuation). Imagine a chef assigned the task of cooking dinner. There are two ways the task can be organised. In the first, the host or client picks out a menu in advance. All the chef needs to do is list the ingredients needed, shop for them, and then actually cook the meal. This is a process of causation. It begins with a given menu and focuses on selecting between effective ways to prepare the meal.*
>
> *In the second case, the host asks the chef to look through the cupboards in the kitchen for possible ingredients and utensils and then cook a meal. Here, the chef has to imagine possible menus based on the given ingredients and utensils, select the menu, and then prepare the meal. It begins with given ingredients and utensils and focuses on preparing one of many possible desirable meals with them.*

EVA RÚN MICHELSEN
Eva was the first CEO of The Ocean Cluster House.

The Ocean Cluster House is like the kitchen, filled with ingredients and utensils; it is up to the entrepreneurs to find the resources available to them and figure out the best mix. This is often how startups evolve. The entrepreneur develops the business idea using the resources available, through relationships and exchange of ideas. In classic business books, this need for relationship and networking early on in the life of the startup is often overlooked. Today, more and more business literature is picking up this important issue.

The Ocean Cluster House is a gathering place for different resources which are linked to the ocean, including fishermen and seafood processing technicians, product designers, marketing and sales people, inventors, social media specialists, biochemists, marine biologists, etc. The entrepreneurs use their relationship skills to observe these resources, exchange and test new ideas on them, and try to form a coalition with those they see fit. In The Ocean Cluster House, there are 70 companies of different sizes which represent most parts of the ocean value chain in Iceland from fisheries to seafood biotech companies. The Cluster management has carefully selected the companies so that they represent a variety of knowledge and skills but still have a common thread. The 65 additional companies which are in the cluster but do not have space in The Ocean Cluster House form an extended network. The cluster management seeks ways to network with startups and others in The Ocean Cluster House.

The challenge of the Cluster is to channel all these resources together so that the entrepreneurs have more desirable business "meals" to test. The key here is a sensible selection of companies and good coffee. Remember that dynamic businesses continuously seek connections, especially in an environment which is created as a community — a place for people to exchange ideas and create value.

...and a coffee machine

At the heart of all of this are two coffee machines. Here, we take inspiration from Google. We learned early on that large tech giants use physical space to bring people and ideas together. The coffee machines play a significant role there. So, as we learned from Google, they did not want each office in the same corridor to have their own coffee machine, but rather use the coffee machines to connect people. We were daring enough to ban companies in The Ocean Cluster House from having their own coffee machines, and it worked. Every Friday morning, the staff within The Ocean Cluster House get together for a coffee, which has become a great part of our weekly routine.

The coffee machines themselves are magnets for interaction, where members meet on a professional as well as a personal level. These machines are a very important part of co-working spaces like The Ocean Cluster House. Instead of each office having its own coffee machine, we get people to connect through two nuclei around the House. It seems like a simple point, and it is, but it makes all the difference. The coffee machines also make hot chocolate, which could be the real secret!

The founder of a sea salt manufacturing startup which had an office in The Ocean Cluster House was standing by the coffee machine, discussing his company's challenges in obtaining salt from the ocean. There were some technical issues which had to be addressed. A very successful founder of 3X, the high tech seafood processing company, happened to be standing in the coffee room. He overheard the conversation, and there was magic! They started talking and after three weeks the problem was solved with help from 3X — actually, no money changed hands!

In another interesting development, The Ocean Cluster House is rapidly becoming a hub for networkers. I believe our mission smittens others and attracts people with similar motives for increasing cooperation. Therefore, today we have several companies which focus on bringing people together, investment networkers, a group of fisheries collaborating on sustainable fish catch certifications, etc.

It is safe to say that, through our coordination and development efforts, we have only begun to scratch the surface of the opportunities and possibilities. This is what progress and development are all about. More importantly, we do all of this using the best that Iceland has to offer: sustainable energy, brilliant ideas, and even more brilliant people.

The Norwegian cluster model

Even though the Iceland Ocean Cluster was formed and financed without government involvement in a society without a clear-cut framework for clusters, I strongly believe cluster success in countries like Norway shows the importance of a national cluster strategy. The Norwegian cluster strategy helps to build a viable cluster community. Norway is at the forefront of cluster development with its government-supported cluster program. The goal is simple: increase cluster dynamics.

The Norwegian cluster program aims to trigger and enhance collaborative development activities in clusters. The goal is both to increase the cluster dynamics and attractiveness, as well as sharpen the individual companies' innovativeness and competitiveness. The cluster program has four major priorities: (1) the general operation and development of the cluster; (2) links between the cluster and the most relevant national or international research, development, innovation and educational institutions; (3) collaborative innovation projects through platforms or infrastructure to identify or develop new products, services or technologies; and, (4) promoting collaboration between clusters around technology, innovation, expertise or business development.

In the Norwegian model, clusters compete to be part of the program. There are strict criteria to take part in the program. The clusters are evaluated on the basis of their strategy for collaboration, the leadership of the industry, and their execution plan.

Excellent work, Norway!

Strategy is key

Morrison (2018) gives a good summary of the steps taken in cluster acceleration. In line with the low hanging fruit ideology, Morrison emphasises the importance of the conversation as the initial tool for networking. This description mirrors quite well our experience with many of the projects which have accelerated within the IOC and found success.

» *The conversation shifts* Innovation ecosystems begin to form with conversations among companies that share a similar 'competitive space'. These conversations typically focus on either common problems or opportunities that could emerge by linking and leveraging assets.

» *A new network forms* As more companies and organisations join these conversations, the connections among individuals become stronger. Participants become aware of an emerging network within the region. One or two organisations emerge as 'network hubs' that start to concentrate shared assets within the network.

» *A strategic agenda emerges* Members of the emerging network begin to focus on strategic opportunities. They learn how to move the network strategically to focus on specific, measurable and pragmatic outcomes. 'Strategic doing' has become an effective discipline to structure and guide these conversations.

» *Participants commit to anchor investments* As the ecosystem forms, members develop a strategic investment agenda: a portfolio of shared investments to accelerate innovation. The portfolio includes larger-scale shared 'anchor investments'. The portfolio includes investments in talent; entrepreneurship and innovation support networks; new narratives to energise and expand the network; and quality, connected places such as incubators and research centers.

» *The cluster emerges as participants continue to invest, adapt and expand*
Connections within the network become more dense and spontaneous. New anchor investments build out the infrastructure of the ecosystem. In addition, new, innovative networks emerge and connect as 'boundary spanning' firms connect with other firms, markets, and opportunities.

In line with Morrison's summary, the opportunities within the IOC emerged when the network became active. For the cluster manager, it means he or she needs to be continuously holding festivals of ideas with large or small groups! Getting people to meet will always be the anchor of the cluster. At the same time, the cluster manager and its leadership must always remember the long-term strategy, even among the many meetings and "low hanging fruits".

The cluster manager is in the best position to see the big picture. In our case, we noticed many small projects forming in the cluster related to the better treatment of raw material.

I believe the reason was that the cross-pollination was new in the network; there were new ideas and very different knowledge — from biotech and consumer goods specialists to fishermen — which opened new doors.

100% in
Numbers

How much value can the 100% utilisation strategy bring to the seafood industry? Our team at Ocean Excellence did some math. If we assume wholesale for a 1 lb. cod is $1.00 immediately upon harvesting, this is our estimation for value added to the price by using the byproducts.

The liver can add 36% to the dollar. Here we assume the liver is developed into an omega-rich liver oil for human consumption. The heads and the bones can add 25% to the dollar. Icelandic cod heads have been dried with geothermal power and exported to Africa.

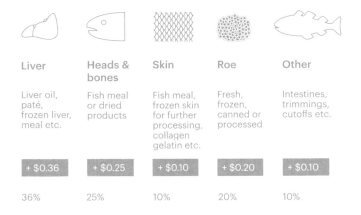

Liver	Heads & bones	Skin	Roe	Other
Liver oil, paté, frozen liver, meal etc.	Fish meal or dried products	Fish meal, frozen skin for further processing, collagen gelatin etc.	Fresh, frozen, canned or processed	Intestines, trimmings, cutoffs etc.
+ $0.36	+ $0.25	+ $0.10	+ $0.20	+ $0.10
36%	25%	10%	20%	10%

We are very conservative in our estimation of the value of the fish skin. We estimate the skin adds 10% to the dollar. The skin is then developed into collagen and sold in bulk. Moving further up the value chain, the fish skin can create value which would much more than exceed the dollar value of the cod. The roes can be canned which gives a 20% increase in the value to the dollar.

Other utilization such as enzymes from the intestines and calcium from the bones, we carefully estimate to be 10% to the dollar. This conservative estimate shows a possible doubling of the value of each fish being caught.

Value addition is, of course, not only applicable to cod; we can do estimate like these to most fish species. Here, it is crucial to find ways to make full utilisation a good business for both fishermen and fisheries. Each needs to benefit.

Our Mission: 100% Fish

From the early days of the cluster, we were often asked to summarize our main task and our mission. Our ongoing work strategy is to "create value for the companies we are working with by using cluster methodology". I have never been a great fan of mission statements. They are often boring and even banal. Often, mission statements are written without involving employees, so they do not stick. How should a cluster come up with an inspiring mission statement, highlighting why people should choose to be a part of our community and agenda? We wanted to involve the cluster members, and the best way to do that meant following them in their cluster work.

After the first two years, we realised that the cluster members were very interested in full utilisation of fish. Five working groups were formed in the cluster focusing on the various new uses of fish byproducts or byproduct processing technology. We also noticed the media was very interested in more utilisation of fish. In the early days of the cluster, I was often told by people in the seafood industry that the media did not show any interest in writing about fisheries. However, as soon as we started introducing our ideas of "100% Fish" and new startups in that field, the media absolutely loved it. We received massive coverage, which helped us to get more entrepreneurs on the bandwagon.

Optimism can be as infectious as pessimism and it can spread rapidly. We felt the cluster should always express optimism and cheerful certainty of

IN COD WE TRUST
*This picture shows the variety of
Icelandic products made from cod.*

the future of the new seafood industry. The more we studied, the more we understood that optimism was well placed in an industry whose core business was to provide the world with traceable, natural and sustainable protein from the ocean.

Even though many academics had been emphasising more utilisation of fish and many projects had been initiated in the '90s and some even earlier,

How to create
～～ a culture for 100% ～～

The list below is a very short guide on how to create a 100% Fish utilisation culture:

Economics

Remind people that the fish byproducts — wild seafood protein from sustainable fisheries — will with time become a more valuable resource! The caught amount of this wild protein is not increasing, but the demand can increase significantly in the coming years. Both fishermen and fisheries need to benefit, particularly as the vessels need to invest in new technology to treat various rest raw materials. This is an issue of economics — a very important issue which fisheries need to address in good cooperation with the fishermen.

Market

Realize the market is different! Existing market, market channels etc, are part of the business decision to discard. To sell skin and body formulas from cod enzymes takes a different team of marketing experts than the ones selling fillets.

Culture

Even though business and markets are important, we still need to enhance a culture for appreciating the opportunities in the new seafood sector. With a greater understanding comes also more appreciation of the research in this area, which is vital to the future of the sector.

A part of a seafood cluster's work is to initiate a public awareness program with a mission to increase popular interest in the new seafood sector. If we are not able to grab the attention of young people, we will not succeed in developing a large industry in the new seafood economy in the future.

full utilisation had not been taken "all the way" to the media and to the general public. We knew that by creating more media coverage about the 100% Fish initiative, we were more likely to enlighten the new generation of entrepreneurs and academics to start working in this field.

Five years later, at least 10 startups focusing on more utilisation of fish had been established. We were making a difference in this field in Iceland. 100% Fish was very much alive.

This grassroots movement, which had been forming in the IOC, gave us an obvious choice when it came to deciding on our 100% Fish utilisation mission: using every part of the fish as efficiently and cleverly as possible. In this way, the cross-pollination of our cluster companies determined our future mission. I believe our continued success of the IOC comes from the mission (and somewhat obsession) about 100% Fish.

Maybe in the years to come we will see more mission-driven clusters. Governments have been encouraging clusters to address particular challenges such as cutting emissions. Maybe incentives from Government will make new clusters more focused.

Most companies have no understanding what the cluster is all about. Talking about creating value through relationships and member get-togethers seldom creates excitement. Most of them are already members of large associations which handle their lobbying, and they don't need to attend more meetings. A clear, game-changing mission for an industry cluster can make a difference here. The 100% Fish mission has done that for the Iceland Ocean Cluster.

Expanding the cluster abroad

I n my early Ph.D. work, I had become very interested in the concept of born globals: companies which, from the beginning, perceive themselves as global. Of course, in such a small country like Iceland, most tech companies who want to expand need to think and act globally. But how was that linked to the cluster, which for most, is a "regional" phenomena? The IOC was not a born global cluster, but very soon we realised our mission of 100% Fish utilisation should be heard outside of Iceland. We know the inherent difficulties in solving a problem that we don't understand. As an example, the massive amount of fish protein used as landfill or discarded into the sea each year in the Western world is a sustainability problem that we must try to understand and ultimately overcome. Iceland has discovered one way of creating value and jobs, especially in remote and rural areas where opportunity is not taken for granted; the 100% Fish initiative. With new partnerships and the sharing of best practices, we may be on the cusp of a new utilisation movement.

When the IOC introduced its work on full utilisation of fish to various foreign sources, a dialogue formed between multiple interested parties around the globe about how to solve the problem of perfectly good fish protein being wasted. There was, and continues to be, an enormous interest in fully utilising resources. The question for us became how best to implement these strategies globally.

Mapping fish waste

To further develop the discussions on fish waste and the possible expansion of the cluster abroad, we went to our mapping strategy. However, instead of mapping companies, we wanted to map fish and fish waste. The economists Haukur Gestsson and Jón Guðjónsson at the IOC wrote a report about the utilisation of cod catches in the North Atlantic Ocean, more specifically Canada, Greenland, Iceland and the Faroe Islands (Gestsson & Guðjónsson, 2012). The aim was to map out an overall picture of how much of the annual cod catch in the North Atlantic is wasted through discards at sea and waste in production. The research relied on data from official institutions located in the countries in question as well as other sources.

The analysis found that there was a statistically relevant difference between the utilisation rates of the different countries. Iceland and the Faroe Islands were ahead of the others when it comes to utilisation, with the highest utilisation rate in Iceland, around 80%. Utilisation in Greenland and Canada were behind the others but were both still on par with each other. Statistics in Norway on cod utilization indicate similar results as Canada.

When the average utilisation rate of all the countries, 54%, is used to approximate the total tons of wasted cod in the North Atlantic, the result is that, on average, 455,000 tons were wasted every year from 1999–2010. This number is, of course, at best an educated guess. It does, however, at least give us some idea of the amount of raw fish material that goes wasted every year in the North Atlantic. As previously mentioned, Icelanders utilise on average 80%, so around 50,000 tons of perfectly good protein from cod is wasted in the Icelandic seafood industry. So there is still an opportunity to do better in Icelandic fisheries.

Opportunities in the North Atlantic

With these results, we realised the IOC could have a larger role to play in wealth creation through clustering in the North Atlantic than we had foreseen in the beginning. Could we attract others with our model and with our culture? We knew our message of more utilisation could bring commercial value to the fishing sectors in the North Atlantic — if only we could get firms to share their knowledge on fish utilisation with each other.

These results became the driving force for the expansion of the cluster network. Once again, we needed to map, and now map the cluster activities in the North Atlantic to see how we could take our full-utilisation ideas further using the existing cluster network.

We believed many opportunities could be created to increase cooperation between the countries, not only on a political level but on a business level as well. By combining forces and sharing know-how, these nations can mutually benefit in manifold ways, forging new opportunities even as they acquire additional strengths when competing in international markets.

Vilhjálmur Árnason, former deputy CEO of the Iceland Vessel Owners' Association, wrote an extensive report on the mapping of the North Atlantic clusters and marine industries, which became a valuable resource.

The report stated:

> The cluster's members can share information about how to increase the utilisation of resources or to reduce waste or environmental impact. It might be difficult for example to increase the total quantity of fish catch, but some opportunities might reveal themselves in relation to the better utilisation of raw materials by collecting information on how to manage the fishing process more efficiently. Such information sharing activities are beneficial for all parties and can only enhance the image of the industry.

> *There is a conspicuous lack of research on the creation of opportunities vis-à-vis natural ocean resources. More extensive and insightful research might open up new avenues of possibilities regarding the production of food for human consumption and feed for aquaculture. The seafood biotechnology sector is still in its infancy, and yet a plethora of opportunities exist that are linked to the development of pharmaceuticals, functional foods, cosmetics, agrichemicals, fine chemicals, proteins, and biofuels.*

The report also emphasised the many opportunities for cooperation in seafood-related affairs between the countries in the north.

> *They may have different strengths and weaknesses, but their cultural ties are strong and there is a robust tradition of working together in many areas for mutually beneficial ends. The main message of this report is that these countries are facing similar challenges and are equipped with unique know how and experience which should enable them to find common solutions.*

But the links between cluster managers will always be different from the links between companies. As discussed in the chapter on low hanging fruits, the initial conversation which starts an idea is not between companies that share a similar 'competitive space', but among cluster managers.

The discussions were still very fruitful, and, in many ways, led to opportunities that could emerge through further networking. I have since discovered that the lack of connectivity was true for most coastal seafood areas, such as in the United States.

The seafood industries in these regions, despite their relative proximity to large and dynamic centers of finance and research, were in essence islands as well: fairly isolated from each other, academia, investors, metropolitan startup hubs, and other various resources.

Future of fisheries
～ and fish processing ～

There are incredible changes in technology and product development in the seafood industry which are shaping this industry right now — creating better and higher margin products, enabling better returns, and saving energy. The most recent and important changes in technology in the seafood industry are seen both on board the ship and in the processing. The new design of fresh fish vessels means the fish goes through a new technological system which allows storing and shipping the fish without ice and at the same time extending shelf life. The capacity of the vessels has increased, and they are more effective. All of this means savings of energy and money.

The second major revolution is in processing on land. Fully-automated processing means that a processing plant in Iceland with an output of 1000 tons would have around 80 employees in 1990, 40 employees in 2010, and, most likely, 10 employees in 2020.

"Waste Into Profit" is the final revolution in the seafood industry. This means we can gain more value from less seafood! On board the newest Icelandic ships, the team is not treating the product as "fish" and "waste". The newest ships have four product streams where the fillets, heads, liver, and intestines are separated from the beginning, making all these products more valuable.

In fisheries and fish processing, people will continue to lose jobs to computers. New jobs in seafood will be created by new seafood enterprises yet to be imagined.

I know we can change this, and we do that by challenging the startup community and the media to get excited about new companies and opportunities in the ocean industry.

Chicken and egg

When clusters are not working, the chicken and egg problem is often the cause. Firms within any particular industry like to be where there are workers, suppliers, and good infrastructure, and workers are willing to move to places where there are good job prospects and opportunities. This is often a real challenge for coastal areas and the seafood industry. Seafood firms need to be close to the ocean resources and workers want to move to places where new jobs are being offered. This seems an ideal match for seafood companies, as the resources are already there, but it isn't. Firms are close to the seafood resources, but the seafood companies are not moving up the value ladder. Half of the industry's resources are wasted and job opportunities in seafood — especially ones demanding new skills such as biotechnology, marketing, design, etc. — are few. This leads to an industry at a standstill.

I have met with many fisheries companies around the world. Finding a common thread which explains why we have this problem of protein waste is difficult and there is not an easy explanation. The problem will not be solved overnight, and solutions must be tailor-made for different fisheries, regions, and geographic areas.

I believe one major reason for the huge waste is the broken value chain. A fisherman in Norway told me that even though he would like to bring remaining raw materials to shore, there was no one at the harbor to receive or process it. Thus, the chicken and egg problem: waste at sea. The chain is broken.

We should note that fish capture is somewhat imperfect, making discards an unavoidable problem to some extent. The degree of utilisation, however, is more manageable. Once caught, spoilage is the biggest enemy of optimal use of the fish: that is, fish decays more rapidly than almost any other food and thus loses its value as a foodstuff. Numerous preservation methods exist — including chilling, freezing, canning, boiling, smoking, drying and salting. Utilisation is thus heavily dependent on how well these processing techniques are managed.

To address the chicken and egg problem, clusters are crucial. The Government has an important role to play. Here, competitive R&D funds are important as they themselves can function as bridges between different parts of industry and academia.

Iceland has built up strong innovation funds which have played a significant role in the advancement up of startups in seafood. Individuals, universities, business enterprises, and public institutions can apply for these grants. Some of these grants focus on technology, others on marketing, still others on research and development.

Individual startup grants are relatively small: often around USD $50–100,000. Still, these funds have made a huge difference. In 2018, ten startup companies in the Ocean Cluster House in Reykjavik received grants from these funds.

One prerequisite for grants is that the entrepreneurs must initiate a cooperative model among R&D, university, existing industry, and startup. By the nature of this relationship, the startup becomes a cluster manager; and the funds force people and businesses to sit down and discuss opportunities, challenges, and new innovations.

One such Icelandic fund focuses solely on seafood innovation. The "Increased Value of Seafood Fund" is relatively small, but having a sharp focus has made this fund successful.

A setup similar to the Icelandic fund model would be crucial for regions which are interested in strengthening the new seafood industry. I believe it could be a game-changer for decreasing waste in the seafood industry and moving toward a "100% Fish" model.

"Good enough
～～～ syndrome" ～～～

In my research, I have noted that Iceland has around 70 technology firms supporting fishing and fish processing because the demand for quality in the fisheries sector is high. Such demand does not come naturally, but as a result of a sensible incentive scheme.

One entrepreneur with a successful cooling technology firm in Iceland told me that, as soon as he had developed his new cooling equipment in the 90's, he visited a large fisheries company in Iceland. He introduced them to his new technology where the temperature management of the fish could, as he emphasised, be improved significantly. The fishermen and the skipper wanted to test the equipment and they later became the first customers.

This same tech entrepreneur took his equipment to other countries, and their response was frequently, "What we have is good enough". There were no incentives to improve quality — fewer incentives to bring a higher standard of fish to the market!

Ocean Cluster Network

Most definitions of clusters share the notion of clusters as localised networks of specialised organisations, whose production processes are closely linked through the exchange of goods, services, or knowledge. The question still remains: how might stronger cooperation or harmonisation among ocean clusters become a vehicle for improved performance for the seafood industry?

Today there is considerable debate regarding a number of issues associated with clusters. One of the questions is whether regional diversity or specialisation can promote knowledge spillovers. Other questions include "How do clusters work differently when they are working on a regional, national, or cross-border basis?" and "What composes a cluster: firms or industries, local or regional factors?". Location can influence how clusters function. In some cases the existence of clusters can rely on the idea that a good deal of competitive advantage lies outside companies and even outside their industries, residing instead in the locations at which their business units are based.

Silicon Valley is a good example of a strong, internationally known regional cluster. But do clusters have to be regional? Porter (2000) points out that changes in technology and competition have diminished many of the traditional roles of location. According to Malmberg and Maskell (2002), empirical studies have not been able to provide evidence that proximity between co-located firms is necessary for the exchange of information and knowledge.

Ketels (2017) discusses global cluster networks:

> *"While cluster research has traditionally focused mostly on local linkages, the global value chain work looks at industry-specific linkages across locations. It responds to the observations that value chains in many industries are increasing 'unbundled' across locations (Baldwin, 2006). The cluster literature views global value chains as a natural addition to specialised local clusters, as the global pipelines that naturally interact with and complement the local buzz. It is a reminder that cluster-based policies should not focus exclusively on strengthening local linkages, but view the local linkages as a way to support a unique positioning in global markets and industries. The value chain literature is instead often more focused on understanding how locations can enter and enhance their position within value chains".*

Bringing different clusters or cluster managers together across borders to form an alliance is not an entirely new idea. It has been done, for example, in the Medicon Valley Alliance, where 270 private and public members from Sweden and Denmark work together. Another example is the Baltic Sea Region Program (BSR), which is a European Union initiative (2007–2013) that promoted regional development through transnational cooperation. A third example of interest is the Northern European Competence Network for Offshore Wind Energy, a transnational cluster led by BIS in Germany together with 17 other partners from Germany, the United Kingdom, Denmark, The Netherlands, Norway, and Sweden. The cluster partnership is built on a range of expertise from offshore wind energy to oil and gas.

Once people have learned to work together and have enjoyed the benefits of their regional cluster, one interesting option is to enlarge the cluster circle and to start working with people in similar clusters in other countries. In 2015, the IOC decided to emphasise the urgency of building a network of clusters which might address the under-utilisation of caught fish on a global scale.

The urgency for a global network

Improving global fisheries and efficient use of discards is vital to reducing poverty and creating vibrant coastal communities in the world. Even though most global institutions have emphasised the importance of sustainable fisheries to feed poor people in developing countries, this challenge is not at all confined to the poor parts of the world.

Our interest in forming a network of clusters was sparked by our visits to various coastal areas in Europe and North America. I was approached by people in charge of waste disposal units of various municipalities who told me one of their biggest challenges was how to get rid of fish waste. Most of it became landfill! They told me the amount of fish waste, both from fisheries and aquaculture, was a significant problem in their operations. "Fish waste" is a term which Icelanders rarely use, as we see all the parts of the fish as protein and a valuable commodity.

Many fishing towns in Europe and North America have become marginalised. There are fishermen, infrastructure, and even good educational facilities, but the startup world and financial institutions show very little interest. These communities are in great need of opportunities, and they hope improving their industry can keep their new generations from leaving. Is there an opportunity here to open a global dialogue and get a movement started?

As I have discussed in this book, there is a disconnect between many parts of the value chain in seafood. As a result, it is a worthy challenge to get this chain to work more efficiently and create value.

How is the
~~~~ IOC financed? ~~~~

Industry clusters are mainly financed through Government funding and membership fees. The Iceland Ocean Cluster has never received government grants for the core operation of the cluster. The IOC has received payment from Government organisations and international R&D funds for specific projects, analysis and advisory work.

Individual company membership fees range from equivalent to USD 1,000–USD 15,000, where the largest firms pay the highest rates and the smaller firms pay less. Startups have never paid any membership fees to the cluster, but many of them have been benefitting greatly from the cluster work. Membership fees account for around 20% of the IOC turnover.

It is difficult to keep all the members of the IOC happy all the time. Clusters are often working on projects which are either benefitting the whole society or a small group of members. Even though clusters try to involve all of the members in "low hanging fruit" projects, it is rare that all of them become involved. Our most valuable members have been large companies with leaders who have been active in the board of the cluster and understand the role of the IOC as an agent of change for the Icelandic ocean economy. At the same time, it is crucial that these leaders encourage their staff to actively look for real projects in the IOC work which can benefit their company directly. Sometimes the projects have benefitted the members and sometimes they have failed. It is crucial to have leaders in these companies who have faith in the cluster management and the overall mission of the cluster, while understanding that clusters can never guarantee success in a project.

Due to the relatively small part which government had played in financing the IOC, the cluster has emphasized on finding other financial means. Early on, the cluster found ways to sell equity in successful startups which the cluster had established or was involved with. Some of the startups, still partly owned by the IOC, have perfomed well and have provided income for the cluster, thereby strengthening the IOC. The Iceland Ocean Cluster is therefore becoming an active business accelerator and business operator which has made the IOC operation much more financially sound.

The Ocean Cluster House has always been operated independently of the IOC. The Ocean Cluster House charges a market rent for all the tenants, large and small. In its short history, The Ocean Cluster House has most often had modest returns on equity.

The first move in expanding the cluster happened as a result of an expansion of one of our cluster members, Eimskip shipping, into the state of Maine in the US. We met a delegation from Maine at The Ocean Cluster House. They had learned about our work regarding the mapping of ocean related clusters in the North Atlantic and were interested in joining our network. We felt very strongly that here were enthusiastic individuals who could become great contacts or possible collaborators. These relationships were to become the first formal sister cluster of the IOC outside Iceland, The New England Ocean Cluster (NEOC), with Patrick Arnold becoming NEOC's co-founder and leader.

The media in Maine, and later in other areas in the US and Europe, showed a significant interest in our message. Headlines included "Fish are food and fashion," "Fishing for Innovation: Maine finds Iceland's got it hooked," "Seafood producers told to think outside the box," and "Can we double the value of fish catch in the world?". We knew there was an interest in our message and there was definitely a need! The U.S. is the fifth largest producer and exporter of seafood in the world and the largest importer.

Despite its global significance, there are still opportunities for enhancements within the U.S. fishing industry, especially with regard to fish utilisation. At least 1.5 million metric tons of fish byproducts are discarded in the U.S. per annum, resulting in a value loss of potentially $655 million (Johannesson & Sigfusson, 2019). By increasing the utilisation of fish and fish byproducts, significant value can be salvaged from the supply chain in the U.S. seafood industry. Our experience with the early success of the New England Ocean Cluster has taught us that even though fish species are different from one region to the next, many industry characteristics remain the same and clusters can learn from each other. In most parts of the world, clusters are understood as geographical clusters, but with the relationship building with New England, we felt strongly that a network of clusters could become very impactful.

The early success of the New England cluster is a result of strong local leadership focused on building relationship ties among seafood entrepreneurs and between entrepreneurs and academia. We are continuously learning more, but also realising that all this work is firmly grounded in fairly simple ideas of human interaction.

Other sister clusters have emerged that we are very excited about. In 2017 we co-founded the New Bedford Ocean Cluster (NBOC) in Massachusetts with Edward Anthes-Washburn. In 2018, we co-founded the Pacific Northwest Ocean Cluster (PNWOC) with Lára Hrönn Pétursdóttir. Both of these cluster founders and managers have done tremendous work to build their network and start the low hanging fruit strategy of their clusters. In 2019, the Alaska Ocean Cluster (AOC) became the fourth ocean cluster in the US to join the Ocean Cluster Network. The Alaska Ocean Cluster was initiated by Bering Sea Fishermen's Association after my presentation on the Iceland Ocean Cluster model at a public meeting in Juneau, Alaska in 2013. The leaders of the AOC are Craig Fleener and Justin Sternberg. At the same meeting in Juneau, I highlighted the opportunities in full utilisation of seafood which inspired Zach Wilkinson, an entrepreneur, to take action in this field. Zach later became co-founder of Tidal Vision in Alaska which is helping compost fish waste.

"The Ocean Cluster Network" is purpose-driven. This network's vision is to create a context for its clusters that bolsters entrepreneurial growth, trains and attracts worldwide talent, and spurs cooperative innovation. We emphasise 100% Fish use, which addresses the United Nations Sustainable Development Goal 14: Life Below Water. The aim of the Life Below Water goal is to enhance conservation and the sustainable use of ocean-based resources. Although these new clusters have major potential, they will still face many challenges ahead. The most challenging obstacle for them is that, in many areas, there is no understanding or support for industry clusters. In fact, the industries themselves lack knowledge of the role of clusters. As we know, if the cluster ideology is not alive and kicking, industry integration with R&D, Universities, finance, and the startup community is missing. The role of large associations of clusters like the TCI Cluster network (a global network of industry clusters) is very important here — they spread the word.

One should also not forget that at the heart of every successful cluster is the willingness of its participants to demonstrate certain flexibility and to communicate and share information. Through cooperation, it is possible to increase innovation, and through innovation comes production efficiency. In most cases, difficulties are associated with players that have little experience in cooperation and that lack trust. We have found a significant difference between regions and countries in this respect. After visiting Nova Scotia, Canada, and Northern Norway, I was very impressed by the level of cooperation which existed among people in the lobster industry in Nova Scotia, and the whitefish industry in Northern Norway. I saw a completely different picture in other areas, even where I didn't expect there to be a cultural difference. It can take some time to build confidence and open dialogue at a strategic level, and there may be deep-seated animosities between stakeholders (known only to them). Major companies, especially MNEs, may feel threatened, and ego clashes between stakeholders can occur. All these obstacles may be overcome by clusters. I believe a network of clusters can also become a support for the cluster managers who are faced with different challenges in their work.

The existing formula for success in the seafood industry has not included a regional harmonisation or strong cooperation among seafood-related firms.

There may be an opportunity here: can harmonisation increase relations among entrepreneurs within the global seafood industry to strengthen their competitiveness? I believe so. One example could be in the field of ocean health.

As plastic pollution increases globally, there is a need to continue to find innovative solutions to reduce the amount of plastic entering the ocean. Eight million tons of plastic leaks into the ocean every year. If the plastic pollution continues, the amount of plastic waste in the ocean would out-weigh the total weight of fish by 2050. The world needs a paradigm shift, and I believe our Ocean Cluster Network can play a role. Our strength lies in the collaboration between the clusters — to share new ideas in this field, find innovators and connect them with investors and industry. Coastal communities need to increase efforts to clean up the ocean; their livelihood depends on maintaining healthy oceans. Ocean clusters can encourage ocean enthusiasts and people at all stages of the seafood value chain to collaborate. We know collaboration and as plastic pollution is becoming a threat to fisheries in all the ocean cluster areas, our network can lead in ocean protection.

Where are opportunities for ocean clusters?

Our focus within the Iceland Ocean Cluster has been seafood and the whole value chain around it. Within our cluster, we have been open to various interesting projects which have not been a our core strength: ocean tourism, fish farming, and algae development, to name few. These ocean-related industries have all benefitted from being a part of our net-work. We call them "side steps" because there is limited knowledge within our cluster in these areas, but they are definitely interesting businesses which we want to learn more about. Here the Ocean Cluster Network may help. By building on different strengths and sharing knowledge within the Ocean Cluster network, magic can happen.

The New Bedford Ocean Cluster is an example here. Through their mapping, they found interest from the startup world in offshore re-newable energy and developing opportunities for traditional marine businesses. Burgeoning tech segments known as Blue Tech and the Internet of Things (IoT) were both at the leading edge of this interest. This is what makes an ocean cluster network so extremely interesting: we may find different strengths and fields of interest in different regions, which is valuable insight for the network. The Iceland Ocean Cluster has already taken note of the interesting projects in IoT which are ongoing within the New Bedford Cluster and it is in many ways affecting our work. Our sister clusters, both the New Bedford Ocean Cluster and the New England Ocean Cluster, have significantly more knowledge in crustacean shellfishing while our strength lies more in whitefish and pelagic fishing. It is crucial to allow the clusters to build on their own strengths while we try to find ways to get them to share knowledge. Thanks to mapping, the strengths and weaknesses within a cluster become more visible. Whatever strength a community has in ocean-related affairs, I believe the cluster ideology can be very beneficial to strengthen further the parts which cluster members believe has a further potential in their community. After visiting various coastal areas around the world — many of which have shown great interest in enlightening their seafood industry — it has helped me to put on the "ocean cluster glasses" to view the potential of building an ocean cluster.

There are six main variables I look for; together, they represent fertile ground for the development of an ocean cluster. These variables are:

- A robust ocean industry or a strong startup field which is a base industry for the community.
- One or more leaders from particular ocean-related companies who are motivated, willing to be a part of the leadership team, and interested in being spokesmen for the cluster.
- The closeness of Universities or R&D which can become feeding machines of talent and ideas for the cluster.
- Institutions or funds which provide grants to startups or company projects in the field.
- Interest from government (local or national) to back up the cluster initiative.
- A cluster manager/founder with networking/cluster skills.

Whether in a coastal community in the US, Europe or Asia, I have found many — but not all — of these elements to be in place. Most often we found interest from the local government institutions or universities to establish a cluster. In most cases, we have areas where there is a large or fairly large seafood industry. Frequently the seafood enterprises are too weak — they may be comprised of many small boaters who have no time or resources to take part in cluster work, or there may be large seafood companies in the area which have their bases elsewhere. These large seafood companies may be interested in assisting, but they are seldom able to provide people for the leadership group, as the leaders are stationed in a different area. This is one of the greatest challenges for the establishment of a seafood ocean cluster. There will not be any cluster without leadership from the industry. I wish I could suggest some ways around this, but I do feel it will always be meaningless to start a cluster without having strong support from the industry. Several universities which we have met have shown interest in the cluster work and would be a perfect fit for an ocean cluster. Many universities lack a startup culture, but that is changing. The ideology of "triple helix interest" is moving in the right direction within the university system.

Government grants are extremely important in the seafood startup world. The reason is first and foremost that very few private investors have knowledge in this field and therefore the seafood startups tend to be like a black hole in the investment world. Europe has a very good competitive grant system through the European Union which has supported many interesting seafood startup projects. It is the same with the Nordic countries and large European countries. The US system of grants seems to be less focused in this area, but various states are providing grants for startups. It is also understandable that it may well be easier to approach grant providers and investors from a different ocean angle other than seafood. This is hard to accept but very likely the truth. All projects which aim to bring high tech solutions to solve environmental problems in the ocean are a part of an important and popular ongoing dialogue, while fisheries are a part of the past. This needs to change and the ocean clusters need to inspire investors to look more in our direction.

Ocean future

The Iceland Ocean Cluster and The Ocean Cluster House serve as a good example how we can succeed in building networks of people who can create exciting opportunities in the seafood sector.

At the beginning of this book, I discussed the reason for the establishment of the cluster: research indicating weak networks within Iceland's seafood sector. In a study which we did three years after the formation of the IOC, we saw that from the time the seafood tech entrepreneurs became actively involved with the industry cluster work, their patterns of relationship networking changed: they were more actively involved in the network. The obvious question here is: have these increased relations in the IOC create some business or are they just numbers which have no significant effect on the businesses?

Once again, we can gladly use numbers and statistics to back up our claim that clusters like the IOC work. As mentioned previously, 70% of the companies in the cluster have collaborated together; the increase in turnover of the companies within The Ocean Cluster House is much higher than industry average (numbers from 2014–2016).

The number of new startups in the IOC, many of which have been discussed in this book, is also a good sign of successful networking among different parts of the seafood value chain.

The ocean is changing

We are entering a new fish wave, a phase where there is more interest in healthy oceans and doing more with less catch. We hope the emphasis on healthy oceans is not coming too late. As a consequence of human activities, the ocean is changing on a global scale. Increases in CO_2 concentration in the atmosphere are leading to global warming. The consequences are increased temperature in the ocean and ocean acidification. These changes have led to large-scale changes in marine species, ecosystems, and their associated indusries.

I asked the world-renowned journalist and writer Mark Kurlansky, author of *Cod: A Biography of the Fish That Changed the World* to comment on the future of our oceans and Iceland's role in it. He replied:

> *"With both the economy and the culture at stake, no one has more reason than Iceland to save cod. And I doubt anyone has fought harder to save it. But now we can long for the days when the answer was just a well managed fishery. It now seems such a simple solution is no longer enough. The horrible reality is that the North Atlantic is losing its ability to feed all its fish. This will particularly impact larger fish such as cod. Climate change means that the ocean is getting warmer and less salty and carrying increased carbon dioxide, which is attracted to oceans.*
>
> *This makes it difficult for shell fish and plankton to grow, which in turn reduces zooplankton and capelin. There is less and less food for cod so they will become fewer and smaller—no matter how well the fishery is managed—unless we change our energy, drive different cars and change the way we live very soon. What can little Iceland do about such an international crisis?*

> *The one advantage of a small country is that it could become a model country, which would not be enough to save the planet, but it could set the example".*

Kurlansky's message cannot be any clearer. Our oceans are becoming hot, sour and breathless. We should keep on finding ways to create new ocean opportunities in coastal towns, but to keep our ocean healthy we need to act now. Bringing together innovators, industries, universities and startups to come up with clever solutions to save the environment and use resources more sustainably: that is the role ocean clusters and ocean activists can take right now.

It's up to the rest of the world

I strongly believe other countries can use Icelandic experience to ignite their natural resource industry. We need to encourage people to collaborate and establish more startups in their field. Clusters can play an important role in this new wave: to connect people with clever ideas for value creation and supporting the environment, to help the established seafood industry use new opportunities, and to lead the way.

Now it's up to the rest of the world to imagine, and then build, a sustainable and value-added world with 100 percent utilisation.

Acknowledgments

The text in this book consists partly of various articles and analysis published by the Iceland Ocean Cluster between 2011–2018. Special thanks to my friend Vilhjálmur J. Árnason who was one of the first to listen to my ideas and later assist me in many of the writings and events of the Iceland Ocean Cluster.

I want to thank my former employees Haukur Már Gestsson, Bjarki Vigfússon and Jón Guðjónsson for their contribution. Their work on the utilisation of fish in the North Atlantic Ocean, 100% Fish, has added significant value to our work and mission. I hope I have not copied too much of their text without proper citation. If so, please note that I would never have been able to write some of this text without bringing in their writings and thoughts. I hope my mentoring was of some guidance.

I want to thank Dominic Nieper for great advice and guidance in the writing process. Jack Whitacre, who was an intern at the Ocean Cluster and Benjamin Currie, an intern from Reykjavik University, were very helpful in the writing of this book. Thank you, Quentin Bates for your friendship and guidance.

The chapter on the economic and fisheries contribution to GDP is based on a text which was co-written with Professor Ragnar Árnason at the University of Iceland. Ragnar has been my mentor in this field and without his leadership and broad knowledge, this work would never have been realised. Thank you Linda Bryndísardóttir and Eva Íris Eyjólfsdóttir for data collection and analytical work regarding the economic analysis.

I want to thank the CEO of the Iceland Ocean Cluster, Berta Daníelsdóttir, for giving me the time and space to work on this book. Her leadership as a CEO of the IOC has been tremendous and a big part of our success. I also want to thank Eva Rún Michelsen former CEO of the IOC for her significant contribution to the cluster work.

My special thanks to some of the leaders and friends in Icelandic industry, R&D and Government who have been a crucial part of our success: Birna Einarsdóttir, Gunnar Már Sigurfinnsson, Sigurður Valtýsson, Hjálmar Sigþórsson, Guðmundur Kristjánsson, Pétur Einarsson, Gísli Gíslason, Gylfi Sigfússon, Helgi Anton Eiriksson, Grímur Sæmundsen, Katrín Pétursdóttir, Gunnþór Ingason, Ásbjorn Gíslason, Jóhann Jónasson, Þorgeir Pálsson, Árni Oddur Þórdarson, Sigurjón Arason, Eggert Benedikt Guðmundsson, Haukur Óskarsson, Sigurður Ingi Jóhannsson, Kristján Skarphéðinsson, Páll Gíslason, Pétur Pálsson, Sigurjón Arason, Gunnar Tómasson, Eirikur Tómasson, Aðalsteinn Ingólfsson, Einar Lárusson, Guðmundur Nikulásson, Þorsteinn Ingi Víglundsson, Kjartan Eiríksson and Marta Jónsdóttir.

I also want to thank the individuals who have been in the forefront of innovation in fish utilisation for a long time, well before I started the IOC: the amazing team at Matís; Sigurjón Arason, professor at the University of Iceland and chief engineer at Matís; and Einar Lárusson at ORA. You have been great mentors.

Thanks to current and previous staff of Íslandsbanki: Hrefna Bachmann, Rúnar Jónsson, Runólfur Benediktsson, Vilhelm Þorsteinsson, and Edda Rut Björnsdottir played a major role in supporting our analytical work and cluster efforts in the early days. They have all been dedicated to our work ever since.

Thanks to all the entrepreneurs and startups who have been tinkering with their ideas in the IOC and The Ocean Cluster House regarding green technology, traceability, artificial intelligence, and fish utilisation, to name a few. They are the best sign of The New Fish Wave!

To my wife, Halldóra Vífilsdóttir: thanks for your patience, love, and especially your cognitive and strategic design thinking, which has had huge influence on all my work.

References

Arenius P. (2002.) Creation of firm-level social capital, its exploitation and the process of early internationalisation. Helsinki Finland: Helsinki University of Technology.

Agnarsson, S. og R. Árnason. (2007). The Role of the Fishing Industry in the Icelandic Economy. Í T. Bjorndal, D.V. Gordon, R Árnason og U.R. Sumaila (eds.) Advances in Fisheries Economics. Blackwell Oxford, UK.

Agnarsson, S. og R. Árnason. (2003). The Role of the Fishing Industry in the Icelandic Economy: A Historical Examination. Hagfræðistofnun W03:07.

Árnason, R. and Agnarsson, S. (2005). Sjávarútvegur sem grunnatvinnuvegur á Íslandi. Reykjavík.

Árnason, R., and Sigfússon, T. (2011). Umfang og horfur í tæknifyrirtækjum í sjávarklasanum. Frumathugun á fjölda tæknifyrirtækja í sjávarklasanum, þróun og horfur. Íslenski sjávarklasinn, Reykjavík.

Árnason, R., and Sigfússon, T. (2012). Þýðing sjávarklasans í íslensku efnahagslífi. Islandsbanki, Reykjavík.

Baldwin, R. (2006), Globalisation: The Great Unbundling(s), Economic Council of Finland, No. 20, pp. 5–47, Government of Finland: Helsinki.

Bryndisardottir, L.B. (2011). Ekki er allt sem sýnist. Mat á þjóðhagslegri arðsemi íslensks sjávarútvegs. BS ritgerð. Hagfræðideild HÍ.

Burt, Ronald S. (1992) Structural holes: The social structure of competition. Cambridge, MA: Harvard University Press.

Chetty S. (2004). On the crest of a wave: the New Zealand boat-building cluster. International Journal of Entrepreneurship and Small Business, 1(3): 313–329.

Cooke, P., Eickelpash, A. and Ffowcs-Williams, I. (2010). From low hanging fruit to strategic growth. International evaluation of Robotdalen, Skåne Food Innovation Network and Uppsala BIO. Vinnova Report.

Coviello, N. E. (2006) The network dynamics of international new ventures, Journal of International Business Studies, 37, 713–731.

Ffowcs-Williams, I. (2012) Cluster Development: The Go-To Handbook. Cluster navigators: New Zealand.

Field, J. (2003). Social Capital. London: Routledge.

Foster, N. (2006). Exports, Growth and the threshold effects in Africa. Journal of Development Studies, 42(6), 1056–1074.

Gestsson, H. & Guðjónsson, J. (2012) 455,000 tons into the dustbin/sea. A statistical analysis of the utilisation of cod in the North Atlantic Ocean. Iceland Ocean Cluster.

GSGislason & Associates Ltd. (2007) Economic Contribution of the Ocean's Sector in British Columbia. Canada/British Columbia.

Granovetter, M. (1985). Economic actions and social structure: The problem of embeddedness. American Journal of Sociology, 91: 481–510.

Granovetter, M. (1973). The Strength of Weak Ties. American Journal of Sociology, 78 (6): 1360–1380.

Gunnarsson, A.F.(2015) https://hadesignmag.is/2015/12/17/1372/?lang=en.

Jack, S. L. (2010). Approaches to studying networks: Implications and outcomes Journal of Business Venturing, 25(1), 120–137.

Johannesson, H. & Sigfusson.T. (2019) 100% Fish. The U.S. Seafood Industry and Utilisation of By-Products. Arctica Finance & Iceland Ocean Cluster.

Johanson, J. & Vahlne, J-E. (2006). Commitment and opportunity development in the internationalisation process: A note on the Uppsala internationalisation process model. Management International Review, 46: 165–178.

Ketels, C. (2017). Cluster Mapping as a Tool for Development Harvard Business School.

Marshall, Alfred (2013). Principles of Economics, eighth edition. Palgrave Classic in economics.

Malmberg, A., & Maskell, P. (2002). The Elusive Concept of Localization Economies: Towards a Knowledge-Based Theory of Spatial Clustering. Environment and Planning A: Economy and Space, 34(3), 429–449. https://doi.org/10.1068/a3457

Maurer, I. & Ebers (2006). M. Dynamics of social capital and their performance implications: Lessons from biotechnology start-ups. Administrative Science Quarterly 51(2): 262–292.

Morrison, Ed. (2018) https://www.fdiintelligence.com/Locations/Americas/USA/How-and-why-clusters-are-made

Parker, Robert & L. Blanchard, Julia & Gardner, Caleb & Green, Bridget & Hartmann, Klaas & Tyedmers, Peter & Watson, Reg. (2018). Fuel use and greenhouse gas emissions of world fisheries. Nature Climate Change. 8. 333–337.

Powell, W.W. & Grodal, S. (2005). Networks of Innovators. In J. Fagerberg, D.C. Mowery, & R.R. Nelson (Eds.) The Oxford Handbook of Innvovation. pp. 56–85.

Porter, Michael E. 1990. The Competitive Advantage of Nation. Free Press, New York.

Porter, Michael E. "Location, Competition, and Economic Development: Local Clusters in a Global Economy." Economic Development Quarterly 14.1 (2000): 15–34. Crossref. Web.

Pyke, F., Becattini, G. and Sengenberger, W. (1990), Industrial Districts and Inter-firm Cooperation in Italy, Geneva, IILS (International Institute for Labour Studies).

Rangen, C., & Food-Hodne, J., Building Innovation Supercluster" (2019). https://www.engage-innovate.com/reports/building-innovation-superclusters/

Roy, N., Árnason, R., and Schrank, W. E. 2009. The identifiaction of economic base industries, with an application to the Newfoundland fishing. Land Economics 85 (4):675–691.

Sarasvathy, Saras D., (2001) Causation and Effectuation: Toward a Theoretical Shift from Economic Inevitability to Entrepreneurial Contingency. Academy of Management Review, Vol. 26, Issue 2, p. 243–263 2001. Available at SSRN: https://ssrn.com/abstract=1505857

Sigfússon, T., Árnason, R., and Morrissey, K. (2013) The economic importance of the Icelandic fisheries cluster—Understanding the role of fisheries in a small economy. Marine Policy 39 (2013) 154–161.

Sigfússon, T., and Chetty, S. (2013). Building international entrepreneurial virtual networks in cyberspace. Journal of World Business, 48(2), 260–270.